SIMPLIFIED CELESTIAL NAVIGATION

SIMPLIFIED CELESTIAL NAVIGATION

MAXWELL KELCH

HAESSNER PUBLISHING INC

Dedicated to . . .
Laura Belle, Marilyn and Rob
. . . a great crew.

Library of Congress Catalog Card No.: 74-33188
ISBN 0-87799-051-4

All text and captions are set in Helios.
All figures were drawn by the author.
Book Design: Randall De Leeuw

Haessner Publishing, Inc.
Newfoundland, New Jersey 07435

Printed in the United States of America

CONTENTS

How to read this book . . . .

Celestial navigation, although a very simple skill, uses language that is new to most people. This is one reason why such a mystery surrounds the subject. To keep from getting "bogged down" in mechanics and nomenclature, the best way to read this book is to sit down and read it from "cover to cover" without trying to work through any of the examples. It will take less than two hours and you'll have a quick "over-view" of the subject. Then go back, re-read the theory that isn't clear to you, and check the meaning of the terms in the glossary if you find it necessary. Afterward, concentrate on the practical examples starting with Chapter 15. Make frequent reference to the Work Sheet accompanying each problem. You'll soon see how easy it is to fill in the figures and obtain a "fix".

Good luck, and have fun!

PREFACE

Over many centuries, an air of mystery has surrounded celestial navigation. People are fascinated with the thought that by observing the positions of the sun, or the stars, or other celestial bodies, their exact position on the surface of the earth can be determined. The person who can convert observations from the heavens to a spot on a chart is credited with almost supernatural ability, or at the very least is considered a mathematical genius.

The literature that has been published on the subject over the years doesn't help to clarify the process for the average boatman. The two outstanding publications in the field, namely, "Dutton's Navigation and Piloting," published by the United States Naval Institute, and "H.O. Publication No. 9, American Practical Navigator - Bowditch," published by the U.S. Naval Oceanographic Office are technical books that would scare most people to death. The many celestial navigation manuals published for boatmen are mostly "rule books." They do not give the reader a basic understanding of the theory of celestial navigation. They do not contribute to a person's knowledge of why he does certain things to obtain a position at sea from celestial observations.

In addition, there are literally scores of tables published and methods used to solve celestial navigation problems. Which method is best is subject to endless debate by individuals.

This book has been designed to make the theory of celestial navigation understandable to this country's average small boat owner. Then, after presenting the theory, the mechanics of "working up a sight," using the simplest method now available, will be presented. Admittedly, other methods will produce theoretically greater accuracy than the one we will use; however, the end result is all that counts.

Celestial navigation "fixes" are based on the ability of an average boatman to obtain a "sight" from the deck of a boat while bouncing around on a rough sea. The accuracy of the resulting "fix," or position, is not determined by the accuracy of the tables or method used, but by the accuracy of the sextant observation. The accuracy resulting from a mathematical process is limited by the accuracy of the "weakest link" in the chain of calculations.

Let's illustrate this point with an example. The circumference of a circle is calculated by multiplying the mathematical value of "pi" by the diameter of the circle. If one measures the diameter of a circle with an ordinary hardware-store yardstick, which obviously isn't accurate to a hundredth of an inch, there is no need to take the value of "pi" out to eight decimal places to make a calculation of the circumference of the circle. Using a value of "pi" as 3.14 will give you results accurate enough to be used with a yardstick measurement. Using "pi" as 3.14159265, which is of course a more accurate value of "pi" will only cause you to do more work, and increase the possibility that you will make a "bonehead" error in arithmetic. In small-boat celestial navigation, the sextant sight is the "yardstick" in the problem.

Thus, in what follows, we are going to use the simplest methods available, namely the use of the Air Almanac and Sight Reduction Tables H.O. No. 249.[1] The only other equipment needed for our work is a sextant, (a plastic $15 model will do to begin with); a protractor or plotter; dividers or compass; a portable shortwave radio and an ordinary stop watch. These supplies, with the possible exception of the radio, are generally obtainable from a marine supply store or ship chandler in major cities and waterfront communities.

In addition to giving the average person the ability to navigate "by the stars," this book will open the pages of Dutton's and Bowditch, and perhaps encourage the further study of the fascinating art of Celestial Navigation.

San Diego Yacht Club
San Diego, California
September 1974

1. Dutton's: "H.O.249... is a satisfactory method for small vessels making long ocean voyages. Stowage space in such ships is limited, and the three volumes give world wide coverage....In accuracy, H.O.249 is compatible with the accuracy of the sextant observations that can be obtained aboard a small craft in rough water."

INTRODUCTION 1

Thousands of years ago man practiced the art of navigation. Although the early mariner's instruments were few and crude, he sailed the unknown oceans of the world. The astrolabe, forerunner of the marine sextant, was invented before Christ, and long before the Christian era astronomers had prepared tables listing the declination of the sun. The "noon sight" used today to determine latitude, was in general use by experienced navigators in the Fifteenth Century. Long before that, the pole star was used as a guide by which to steer; and by the Sixteenth Century the method of determining latitude by Polaris came into general use among seamen.

Captain Bligh and eighteen men, set adrift in a twenty-three-foot open launch by the "Bounty" mutineers in 1789, navigated safely across thirty-six hundred miles of uncharted South Seas to make their chosen landfall on the Island of Timor. Captain Bligh could navigate the seas accurately then. How much easier it is today. In 1802, Nathaniel Bowditch published the first edition of "The New Practical Navigator," (now H.O. Publication No. 9, "American Practical Navigator"). Today, some three quarters of a million copies later, it is still the final authority for the generations of navigators who came after him. Bowditch's simplified methods, which he vowed to be able to teach to any average seaman, paved the way for modern navigational methods.

Celestial navigation is fun, and it is simple. Anyone who can take a sight with a sextant, add and subtract a few simple figures, and refer to a simple set of tables, can be a good navigator today.

The availability of the modern transistorized portable radio, used to receive "time-ticks" that are broadcast continuously, eliminates the old bugaboo of "time." The Air Almanac and Sight Reduction Tables H.O. No. 249 eliminate the old mathematical tedium of "working up a fix."

So let's read on, and soon we will join that mysterious and wonderful group of people known as Celestial Navigators!

CELESTIAL VOCABULARY AND THE UNIVERSE 2

It is possible to write a series of rules for navigators to follow in obtaining a "fix." But if this were done, you wouldn't know anything about the earth and the heavens. You would always be a slave to the rules, much more subject to error, and would not understand what you were doing. It's far better and easier to learn a little about the universe and some celestial navigation vocabulary. Then you will understand the reasoning behind every step as you "plot your fix."

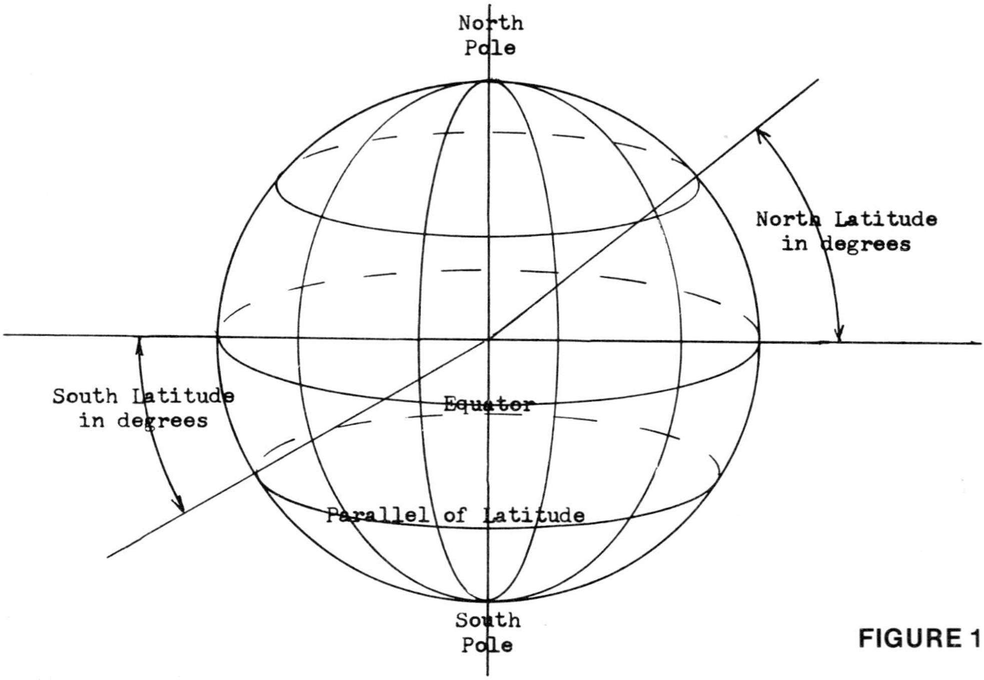

FIGURE 1

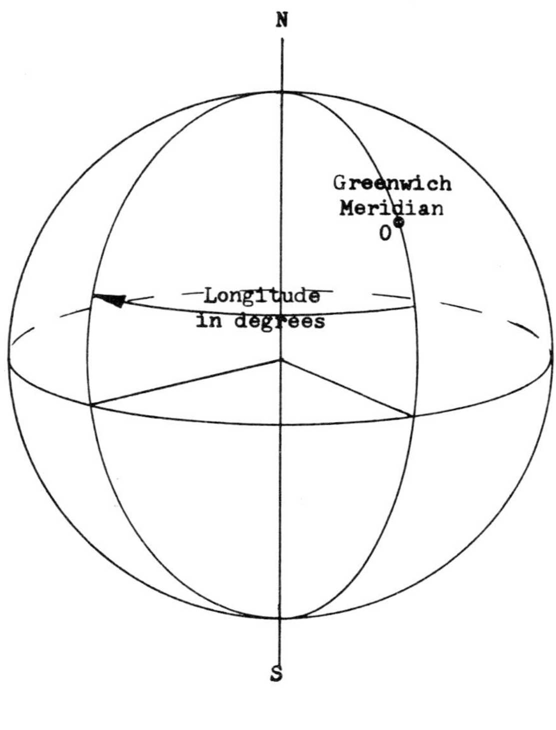

FIGURE 2

THE EARTH 3

The earth is round; not perfectly so, but for practical celestial navigation we can assume it to be a perfect sphere (FIG. 1). The earth rotates about its axis. The axis meets the surface of the earth at the North Pole and at the South Pole. Halfway between the poles is the equator. A plane passed through the equator is perpendicular to the earth's axis. Other planes passed through the earth parallel to the plane of the equator meet the surface of the earth in circles, which are known as parallels of latitude. North Latitude circles are north of the equator. South Latitude circles are south of the equator. The angle in degrees, at the center of the earth, between the equator and a circle of latitude, measures the latitude. Latitude is thus measured from 0 degrees at the equator, to 90 degrees, both north and south of the equator, at the poles.

If a plane is passed through the earth, with the poles and the axis lying in that plane, the circle where the plane meets the earth's surface is called a meridian (FIG. 2).

Meridians measure longitude. There are 360 degrees of longitude around the earth. By international agreement, the starting point for longitude is the meridian of Greenwich; the original location of the Royal Observatory in England. Thus, Greenwich is at 0 degrees longitude. All other meridians of longitude are measured in degrees from 0 to 180 either west or east of Greenwich.

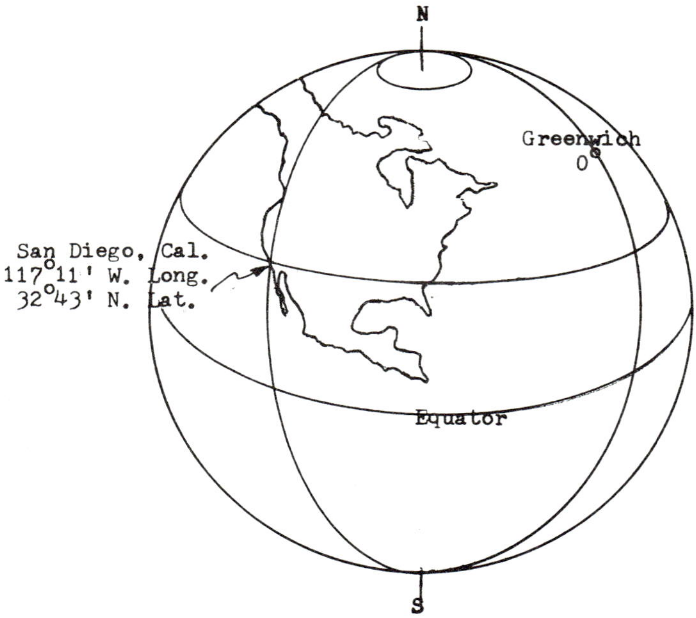

FIGURE 3

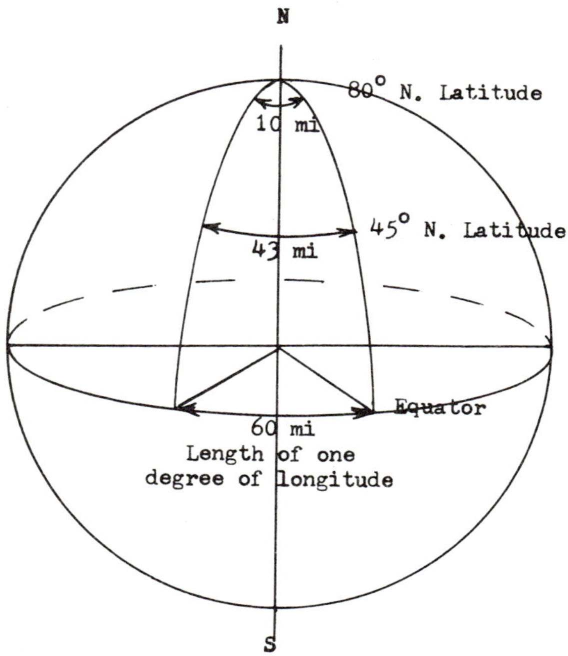

FIGURE 4

With the parallels of latitude and the meridians of longitude, any point on the surface of the earth can be designated. San Diego, California, for example, is located at 32°43′ North Latitude, 117°11′ West Longitude (FIG. 3).

The meridians of longitude cut segments out of the earth that look like the segments of an orange (FIG. 4). One degree of longitude along the 0° parallel of latitude, the equator, measures 60 nautical miles on the surface of the earth. The distance of one degree at the equator, Latitude 0°, is greater than along a degree of longitude near the poles, Latitude 90°. At the poles the longitude distance actually is zero. The surface of the "orange segment" gets smaller as you approach the poles.

The distance along a meridian, for equal degrees of latitude, is the same everywhere, since all meridians are "great circles." A "great circle" is the circle traced on the surface of the earth by a plane that passes through the center of the earth. The equator is the only "great circle" of latitude. The latitude circles get progressively smaller as the parallels of latitude approach the poles, since they do not pass through the center of the earth.

Since there are 60 minutes in a degree, one minute of angle on a great circle, measured from the center of the earth, is one nautical mile on the earth's surface. A nautical mile is 6,076 feet.

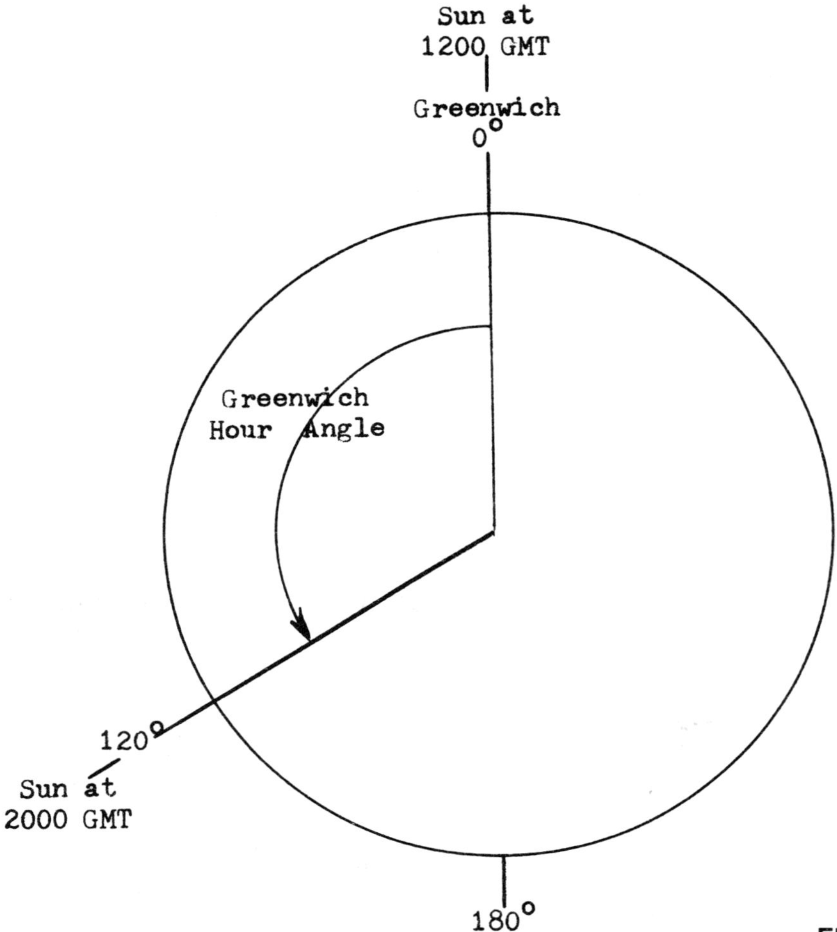
Sun at
1200 GMT
Greenwich
0°
Greenwich
Hour Angle
120°
Sun at
2000 GMT
180°

FIGURE 5

TIME 4

Now let us put the sun in the sky and find out something about time! As the earth rotates slowly from west to east, an observer on the surface of the earth observes the sun rise in the east and move upward at an angle. At its highest point the sun is on the meridian of the observer, and after it crosses the meridian, at which time it is noon, it starts going down and finally sets in the west. The time from a particular noon to the next is one day, or 24 hours. Thus, the sun completely circles the earth, or travels 360 degrees around the earth, in a day. Every hour the sun travels 15 degrees to the west. (15 degrees multiplied by 24 hours equals 360 degrees, for one day.) If the sun were on the meridian at Greenwich, or at 0° Longitude at noon, one hour later it would be on a meridian 15 degrees west of Greenwich or at 15 degrees West Longitude. If the sun were on the meridian of Greenwich at noon, or 1200 Greenwich Mean Time (GMT), 8 hours later or 2000 GMT, the sun would be on the meridian at 120 degrees West Longitude (FIG. 5). But the time at 120 degrees West Longitude, when the sun crossed the meridian, would be noon, local time. The angular distance between the meridian of Greenwich and the meridian at which the sun is located, is known as the Greenwich Hour Angle of the sun. Thus when the sun crossed the meridian at 120 degrees West Longitude, the Greenwich Hour Angle (GHA) of the sun was 120 degrees. GHA is always measured west from Greenwich through 360 degrees.

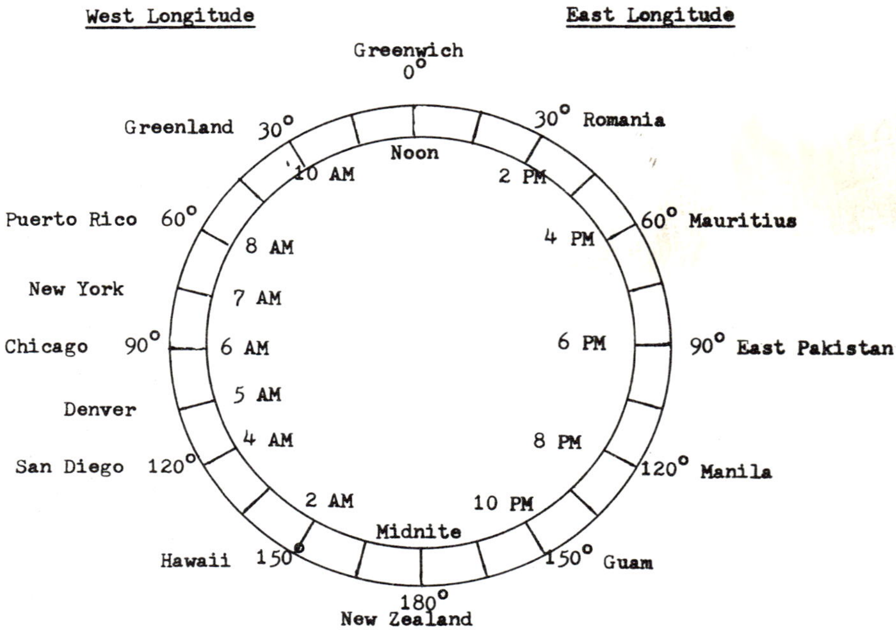
West Longitude
East Longitude
Greenwich
0°
Greenland 30°
30° Romania
Noon
10 AM
2 PM
Puerto Rico 60°
60° Mauritius
8 AM
4 PM
New York
7 AM
Chicago 90°
6 AM
6 PM
90° East Pakistan
Denver
5 AM
4 AM
8 PM
San Diego 120°
120° Manila
2 AM
10 PM
Midnite
Hawaii 150°
150° Guam
180°
New Zealand

FIGURE 6

Time can be equated to the angle through which the earth turns in a given period of time. The time interval between the meridian passage of the sun, between two points on the earth, can be equated to the difference in degrees, or the difference in longitude between the two points.

Since most of us like to get up about the time the sun rises, say around 6 o'clock in the morning, have lunch about the time the sun is overhead, and "call it a day" sometime after the sun sets in the evening, there is a practical necessity of setting our clocks to some local time other than Greenwich Mean Time, which is the local time in England.

As a result of this need, Zone Time has been established throughout the world. In Zone Time, all clocks in a given time zone, the centers of which time zones are roughly 15°, or one hour, apart, are set at the same time. Clocks are reset when one moves into an adjoining time zone. The Eastern Time Zone, 75° west of Greenwich, is 5 hours earlier than Greenwich; while the Pacific Time Zone, 120° west of Greenwich, is 8 hours earlier than GMT (FIG. 6).

Eastern Standard Time is found by subtracting 5 hours from GMT. When it is 1200 GMT or noon at Greenwich, it is 0700 or 7 AM in New York. Conversely, if it is 0700 EST in New York, you have to add 5 hours to get 1200 GMT in England.

Daylight Saving Time requires an adjustment of one hour. In the Spring of the year, clocks "spring" forward by one hour; and in the Fall of the year, they "fall" back by an hour in zones operating on Daylight Saving Time. In other words, during Daylight Saving Time, the local time zone is changed to the Standard Time of the zone adjoining to the east. Pacific Daylight Time is the same as Mountain Standard Time.

If you could leave Greenwich at noon on a particular day, in a very fast airplane, and fly westerly keeping the sun overhead, you would fly around the world with the sun overhead the entire distance. You would arrive back at Greenwich at noon. But the noon of your arrival would be a day later than when you left. The change of date is at the 180th meridian, or at the International Date Line in the Pacific. Crossing from east to west, the date is advanced, while crossing from west to east the date is reduced by one day.

It is important for the navigator to keep the International Date Line, or the 180th meridian, in mind, since a new navigational day starts when the sun crosses the 180th meridian, and a new date is then in effect.

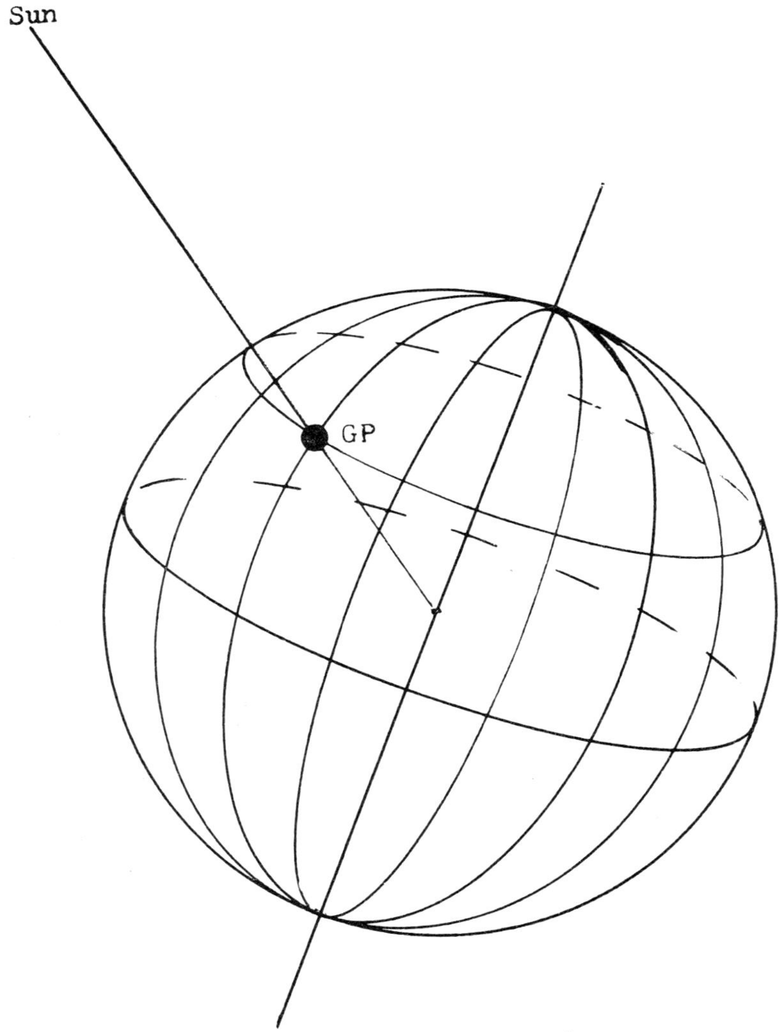

FIGURE 7

GEOGRAPHICAL POSITION 5

If a line were drawn from the center of the earth to the center of the sun, it would pass through the surface of the earth at a point. This point is known as the Geographical Position, or GP, of the sun at that particular instant of time (FIG. 7). To a person standing on the GP, the sun would be directly overhead, or at the Zenith.

Every heavenly body, the sun, the moon, the planets and every star, has a GP, and this GP can be determined in advance for any instant of time, for any day, for any year, by astronomers.

The GP of the sun moves across the surface of the earth at the rate of 15 degrees per hour, of 15° x 60 = 900 angular minutes per hour, which is 900 nautical miles per hour at the equator, or 900 knots. In one minute of time, the GP of the sun travels 15 nautical miles. Or in every 4 seconds of time the GP of the sun moves one nautical mile to the west.

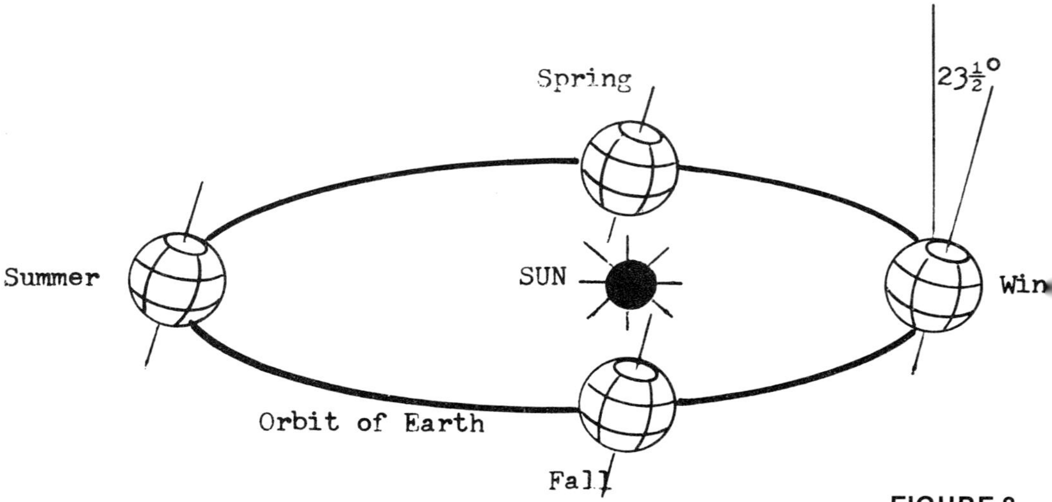

FIGURE 8

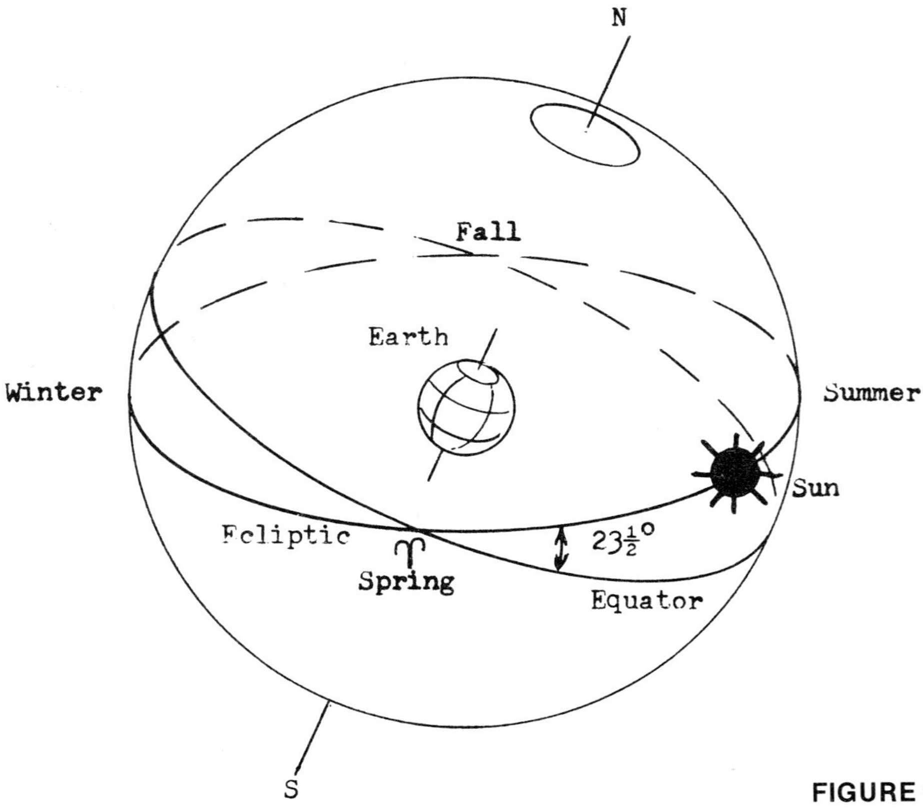

FIGURE 9

THE EARTH'S ORBIT 6

The sun is the center of our solar system and the earth and the other planets travel in orbits around the sun. The sun itself is in motion in space. In the case of the earth, it takes a year for it to make a complete circuit around the sun along an elliptical path. Since space has no limits, no fixed boundaries, no range markers, no up and no down, no beginning and no end, the motion of any object in space must be described as it relates to some other object or position in space.

There are two ways to describe the motion of objects. One is to stand in some motionless spot and describe the motion of everything you see relative to that spot. This is called "actual motion." The other way is to describe the motion of everything you see from a position on one of the moving objects. This is called "relative motion" or "apparent motion."

Most of us are self-centered, and so it is with celestial navigation. Our vantage point is on the earth and, as we look out into space, we assume that the earth is the center of the universe, and we describe everything as we see it from earth. All of the heavenly bodies are located on a celestial sphere with the center of the earth as the center of everything. We also assume that all heavenly bodies, with the exception of the moon (which is very close to us astronomically speaking) are so far away that they are all located on the celestial sphere. In other words all of the heavenly bodies, except the moon, are so far away that, for navigation purposes, the light from them comes to the earth at the same angle, whether viewed from the North Pole, the South Pole, or from any other spot on the earth in between. We consider then that all heavenly bodies, except the moon, are located at infinity on the surface of the big blue bowl of the heavens, the celestial sphere.

As the earth rotates on its axis from west to east, the celestial sphere appears to rotate. The stars on the celestial sphere rotate from east to west, rising in the east, crossing the meridian of the observer, and then setting in the west. Some stars, sufficiently high in the sky, never rise nor set, but rotate around the celestial pole. These stars are called circumpolar stars. In the Northern Hemisphere they rotate counter-clockwise around the pole star Polaris.

For our navigation purposes, all of the stars on the celestial sphere are considered fixed in relation to each other. In other words, they do not move relative to each other.

Probably the earliest recognized of all astronomical phenomena was the sun's apparent annual motion among the stars. In the summer the constellations rising on the horizon at sunrise are not the same as the constellations rising at sunrise in the winter. Also, for the observer in the Northern Hemisphere, the sun rises higher in the sky each day until the middle of summer. Then it starts down until mid-winter when it reaches its lowest position in the sky. It then returns and goes through its cycle again, shifting north and south by 47 degrees in the course of the year. Days are longer than nights during the "summer" half of the year, while the nights are longer than the days during the "winter" half of the year.

The apparent path of the sun, as it moves around the earth on the celestial sphere, is called the ecliptic. The axis of the earth is tilted at an angle of 23½ degrees from the perpendicular to the plane of the earth's path (FIG. 8).

Since we are considering the center of the earth as the center of the universe, or the center of the celestial sphere, let's establish some reference points on the celestial sphere.

First of all, we extend the axis of the earth outward, and where it meets the celestial sphere we have the north and south celestial poles. Then we extend the plane of the earth's equator outward, and its intersection with the celestial sphere marks the celestial equator.

Since the axis of the earth is tilted at an angle of 23½ degrees to the perpendicular of the earth's path around the sun, the ecliptic is tilted at 23½ degrees to the celestial equator (FIG. 9).

Because of this "tilt," the sun is above the horizon for different lengths of time throughout the year. The sun appears to move north and south during the year and the seasons are a result of this motion of the sun.

The sun is on the celestial equator on March 20, which point is called the Vernal Equinox. "Equinox" means "equal night" and the length of the night equals the length of the day on this date. The sun then becomes higher in the sky each day until it reaches a point 23½ degrees higher than it was on March 20. This position is called the Summer Solstice and it occurs on June 21. "Solstice" means "sun standing still." The sun has reached its maximum position and then returns in the other direction. The sun then becomes lower each day, reaching the celestial equator on September 23, the Autumnal Equinox. It continues on down another 23½ degrees until the Winter Solstice on December 22, from which point it starts to rise until it reaches the Vernal Equinox again.

To demonstrate the phenomena that occur as the earth moves along its path around the sun, place a light in the middle of a round table to simulate the sun. Then, holding a ball or globe as a model of the earth, move the globe around the table, while you rotate the globe on its axis, inclined 23½ degrees to the vertical. What we have discussed will become perfectly clear.

The earth travels along an elliptical path around the sun, and although it is closer to the sun in the winter time (Northern Hemisphere), the temperature of the earth is determined by the length of daylight and the angle of the sun above the horizon, not by its distance from the sun.

Since we are closer to the sun in the winter than in the summer, the sun's apparent diameter is greater in the winter than in the summer. The apparent diameter of the sun thus changes by a small amount from day to day throughout the year.

The further north you go in the summer, the longer the days are, and in the region within 23½ degrees of the North Pole, for example, the sun never sets in the summer. This explains the "Midnight Sun."

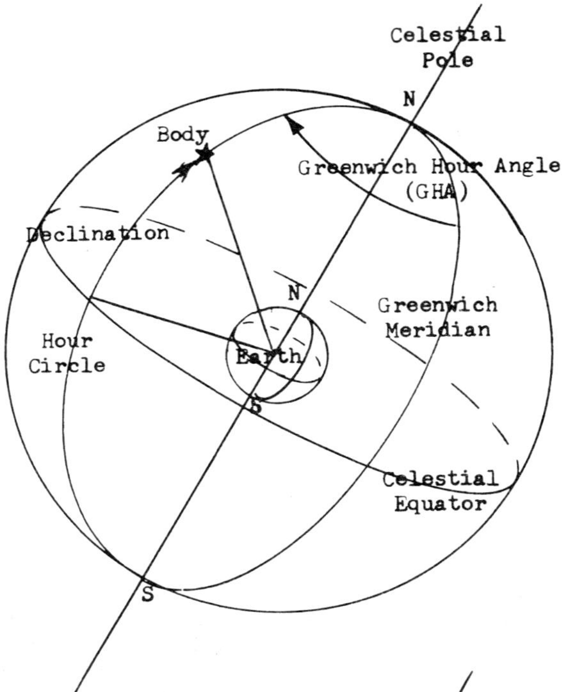

FIGURE 10

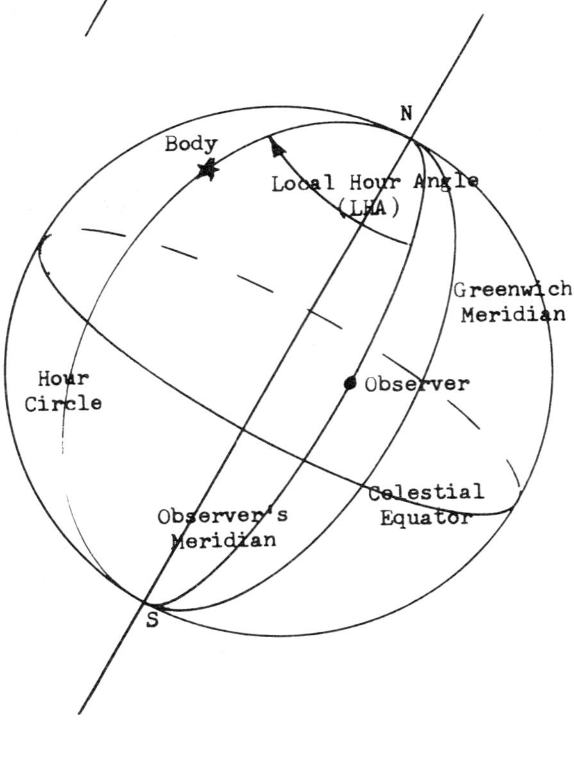

FIGURE 11

THE CELESTIAL SPHERE 7

We said in Chapter 6 that for the purposes of celestial navigation we consider the earth as the center of the celestial sphere. To locate heavenly bodies on the celestial sphere, we are going to use the same methods we use to locate geographical positions on the earth.

We have extended the axis of the earth outward to mark the celestial poles, and the extended plane of the equator has marked the celestial equator. The meridians of the earth can be extended outward, and where they intersect with the celestial sphere they form celestial meridians.

Any spot on the surface of the earth, we learned earlier, can be exactly located in terms of latitude and longitude. Likewise, any heavenly body can be exactly located on the celestial sphere in terms of celestial latitude and longitude. However, on the celestial sphere we change the names. Latitude is called Declination, and longitude is called Hour Angle. The Greenwich Hour Angle (GHA) tells us the number of degrees a body lies west of the meridian of Greenwich. GHA is measured from 0 to 360 degrees **westerly** from Greenwich. Declination is measured both north and south of the celestial equator, from 0 to 90 degrees (FIG. 10).

There are other "hour angles" used in celestial navigation besides GHA, and we should know about some of them.

The meridian that passes through a celestial body is called the Hour Circle of the body. The meridian that passes through the observer is called the Observer's Meridian. The angle between the observer's meridian and the hour circle of the body is called the Local Hour Angle (LHA) of the body. LHA is measured in degrees and it is measured **westerly** from the observer's meridian to the hour circle of the body (FIG. 11).

On the earth, the meridian of Greenwich has been fixed as the reference point from which all **terrestrial** measurements of longitude are made.

In the heavens, a fixed point in space on the celestial equator called the first point of Aries, or just plain Aries, designated (♈), is the celestial reference point (FIG. 12). Aries is the point on the celestial equator where the ecliptic crosses (the Vernal Equinox) as the sun moves from south to north in the spring of the year. Aries is located in the constellation of Pisces (Fishes). The nearest bright star to Aries, or to the 0° celestial meridian, is Alpheratz. Alpheratz, however, is about 30° north of Aries, which is, of course, on the celestial equator.

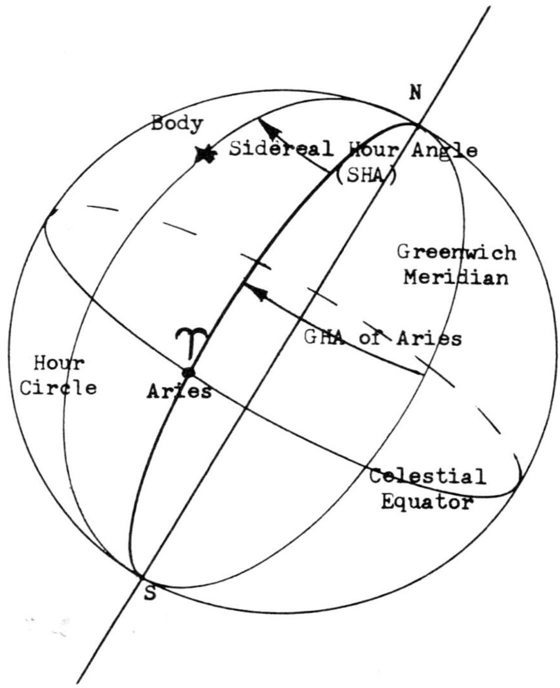

FIGURE 12

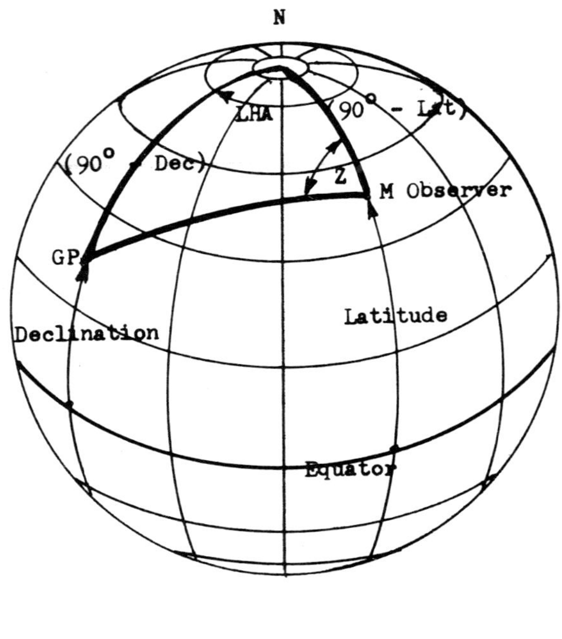

FIGURE 13

The angular distance of a celestial body from the hour circle of Aries is called the Sidereal Hour Angle (SHA) of the body. Sidereal means of, or relating to, the stars. SHA is also measured westerly from Aries, from 0 degrees through 360 degrees.

All stars in the heavens, together with Aries, rotate as a unit. Thus the angular distance of each star from Aries is always the same. The SHA of a star is constant.

The position of any star in the heavens can be determined by its declination and SHA. This corresponds to earthly latitude and longitude. Aries is the reference point for SHA. The celestial equator is the reference plane for declination.

Since the earth is revolving around its axis, Aries is always in motion, just as the sun is in motion. The GHA of Aries, a fixed point in space, is constantly changing, just as the GHA of the sun is constantly changing. As a result, the GHA of all stars is constantly changing. But since the SHA of each star is constant in its relation to Aries, the GHA of a star can be determined by adding the GHA of Aries to the SHA of the star.

As was mentioned earlier, astronomers can determine in advance the exact location in space of any heavenly body, including Aries, for any instant of time, for any day, and for any year. This information is tabulated in almanacs.

Before we proceed to the final system of coordinates needed in celestial navigation, let's look at a drawing of the earth (FIG. 13).

The North Pole is shown, the equator is drawn in, and there is an indication of the parallels of latitude and meridians of longitude. The observer is located at M on his meridian of longitude, and the geographical position of the body is located at GP. Remember that the Declination of a body corresponds to latitude, and Hour Angle corresponds to longitude.

Great circles are drawn, making a triangle with the observer and the North Pole forming one side, the GP of the body and the North Pole another side, and the third side is drawn between the observer at M and the body at GP.

Now what do we know about the spherical triangle that is shown in Figure 13?

The great circle between the observer at M and the North Pole is part of a meridian of longitude. The great circle passing between the North Pole and the GP of the body is on the hour circle of the body. This being the case, the angle between these two circles is the Local Hour Angle of the body, or the angular distance between the observer and the body.

The angular distance of the observer M above the equator is the latitude of the observer; therefore, the distance between the observer at M and the North Pole is 90° minus the latitude.

The angular distance of the GP above the equator is the declination of the body, and thus the distance between the body and the North Pole is 90° minus the declination.

The angle between the observer's meridian, and the great circle between the observer and the GP of the body, is called the Azimuth Angle (Z), and we will talk about this important angle later.

The triangle that we have drawn on the surface of the earth is the Navigation Triangle. In the next chapter we will relate this triangle to the observer as he views things from the surface of the earth.

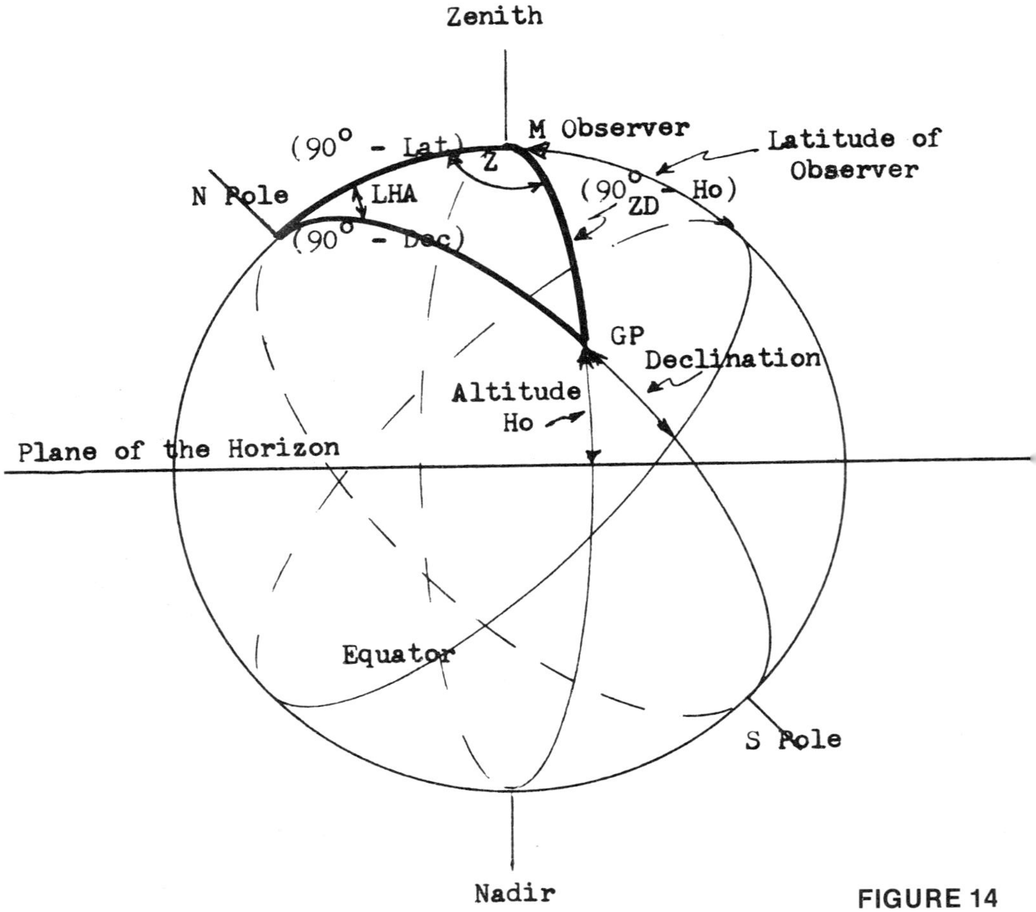
Zenith
M Observer
(90° - Lat)
Latitude of Observer
Z
N Pole
LHA
(90° - Ho)
ZD
(90° - Dec)
GP
Declination
Altitude Ho
Plane of the Horizon
Equator
S Pole
Nadir

FIGURE 14

THE HORIZON SYSTEM 8

So far in our study of celestial navigation we have learned to locate any spot on the surface of the earth by means of latitude and longitude. Then we found that any heavenly body on the celestial sphere can be located using the celestial coordinates of declination and hour angle. Now, the last system of coordinates we need to know something about is the one that moves about the surface of the earth with us as we move about the surface of the earth. This system of coordinates is called the Horizon System.

The Horizon System has as its base, the plane of the horizon; properly, the plane that passes through the center of the earth, parallel to the observer's horizon.

The spot in the heavens, on the celestial sphere, directly over the observer's head, is called the Zenith (FIG. 14). The spot on the celestial sphere directly underneath the observer, through the center of the earth and down to the celestial sphere on the bottom side, is the Nadir. Thus the axis of the Horizon System passes from the Zenith, through the observer, through the center of the earth, to the Nadir.

The height of a celestial body, or the angular distance from the horizon to the body, measured along a great circle, is known as the Altitude (Ho). The angular distance of a body from the Zenith is known as the Zenith Distance. Since the distance from the horizon to the zenith is 90 degrees, the Zenith Distance is always 90 degrees minus the altitude of a body.

Now let's look at Figure 14. This is a drawing of the earth similar to Figure 13. However, since the observer always feels he is "on top of the world," we have drawn the observer's position M on top of the sphere. The Zenith of the observer is overhead, the Nadir is at the bottom, the Poles and the Equator are indicated, and the Horizon is drawn in. The spherical navigational triangle is also shown as it was in Figure 13.

By adding the Horizon System to the drawing, what more do we learn about the navigational triangle?

The altitude of the heavenly body (Ho) is the angular distance of the body above the horizon, as measured along the great circle connecting the body at GP with the observer at M. This being the case, then the distance between GP and M is 90° minus Ho, or the Zenith Distance.

So we have now determined the length of another side of the navigational triangle from a measurement that we can make with a sextant from our boat.

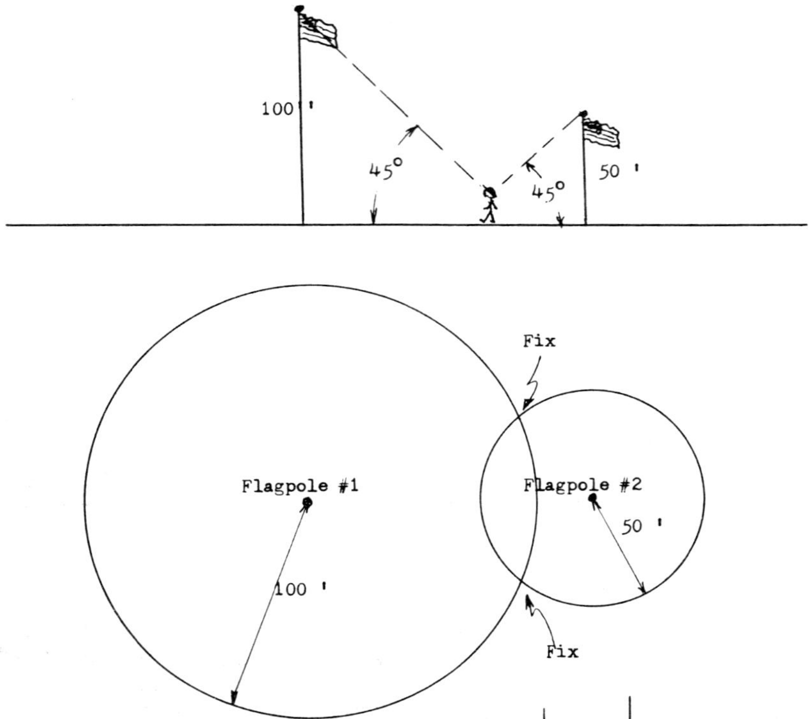

FIGURE 15

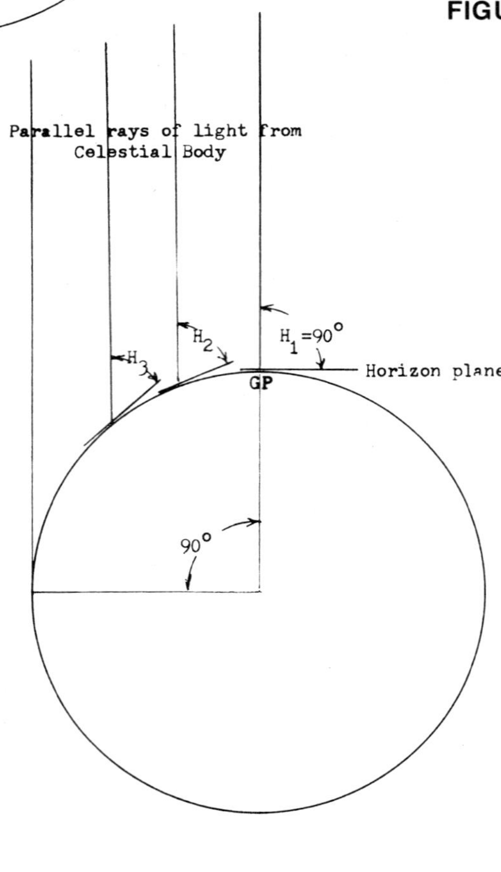

FIGURE 16

LINES OF POSITION 9

Modern celestial navigation involves determining one's position through observations of various heavenly bodies. Observations are made by means of a sextant, an optical instrument that enables the observer to measure the altitude of heavenly bodies in degrees, minutes, and seconds, above the horizon. From the angles observed and the times at which the observations are made, so-called "Lines of Position" (LOP), or "Circles of Position," can be plotted on a chart. Where these lines of position meet, or intersect, determines the location of the observer.

When expressed in the language of the navigator, all of this sounds complicated, but really it is not. All of us use the height of objects to determine where we are.

Suppose, for example, you are visiting in a strange town. When you go downtown on a shopping trip you notice that the City Hall is in the center of town, and a statue on the dome of the building is visible all over the city. The base of the building is the "geographical position" of your reference point. No matter where you are you can look for the statue, and from its apparent height, or angle of altitude, you can determine your distance from City Hall. If the altitude of the statue is great, you are close by. If the angle is small, you are a greater distance away.

Let's take another example. Suppose you are in the middle of a big parking lot in a shopping center. In the center of the parking lot is a tall flagpole, say 100 feet high. You look up at the top of the flagpole and estimate that the angle you are looking up is 45°. Then you know that you are 100 feet away from the base of the pole. As you walk around the parking lot, at any point where the angle to the top of the pole is 45°, you know that you are 100 feet away from the base of the pole. In other words, the angle of 45° from you to the top of the pole establishes a "circle of position" or a "line of position" 100 feet from the base, or GP, of the pole. When you walk closer to the pole, the angle becomes greater; when you walk further away, the angle becomes less.

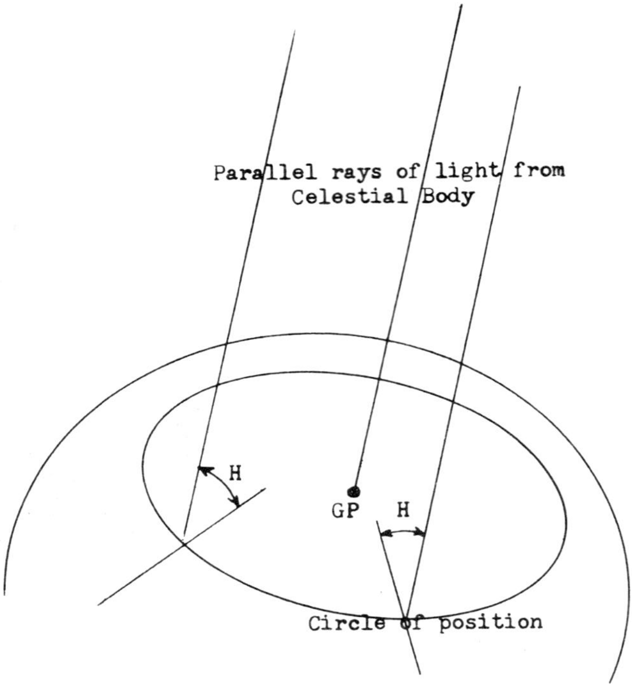

FIGURE 17

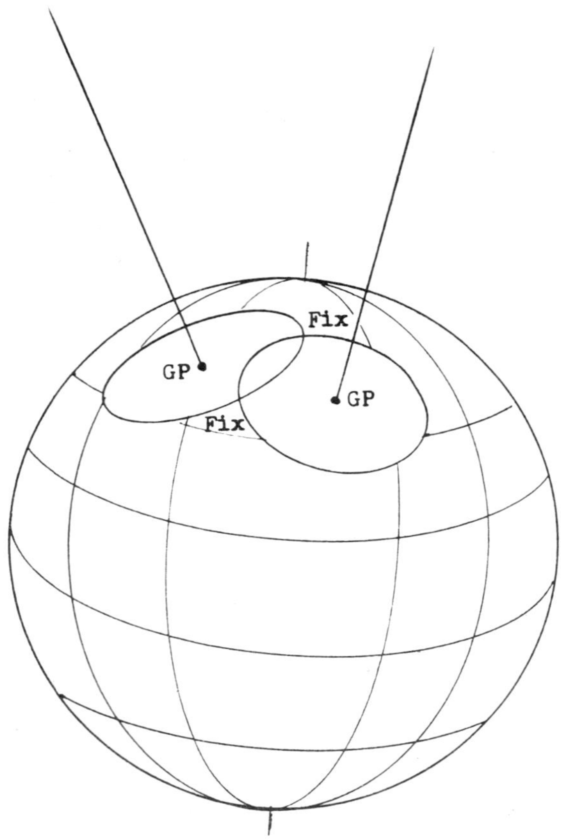

FIGURE 18

If there were two flagpoles upon which you could make observations, you could obtain a second LOP, and the points where the two circles intersect would fix your position (FIG. 15). There would be two placed where the two circles of position cross, so you would be at one position or the other. In celestial navigation, these two points of intersection never cause confusion. The circles of position are so large that one intersecting point is thousands of miles away from your dead reckoning position.

But, you say, this example concerns an object that is very close to the surface of a flat parking lot. How does this apply to heavenly bodies an infinite distance from a spherical earth?

Let's draw a meridian of the earth. We are standing on this meridian exactly on the GP of a heavenly body (FIG. 16). The light from this body reaches the earth in parallel rays, since for practical purposes the body is an infinite distance away. If we are at GP, Position #1, the body is at our Zenith and the altitude H_1 of the body is 90°. If we move from Position #1, the body becomes lower in the sky and the altitude H_2 is less than 90°. If we move farther away to Position #3, then H_3 is less than H_2. As we move away from the GP of a body, the plane of the horizon upon which we are standing changes. We obtain a lesser altitude angle the farther away we move. When we reach a position on the earth 90° from the GP, the body is on the horizon. It has an altitude of 0°.

In Chapter 3 we learned that one minute of arc corresponds to one nautical mile. By moving from the GP to a point where the body is on the horizon, we have moved 90°, or 90° x 60 = 5400′ or 5400 nautical miles around the surface of the earth.

Obviously, we could move in any direction away from the GP of the body, and at all points on the earth the same distance from the GP, the altitude of the body would be the same. Thus "circles of position," based on the altitude of the body, are traced on the surface of the earth (FIG. 17).

Observations of the altitude of another heavenly body, with a different GP, would determine another "circle of position." The intersection of the two circles would "fix" the position of the observer (FIG. 18).

So it now appears that all we have to do to determine our position on the surface of the earth is to make observations of the altitude of two heavenly bodies; look up their GPs in an almanac for the time of the observations; plot the GPs on a chart; and then draw two "circles of position" on the chart. The radius of each circle in nautical miles corresponds to the Zenith Distance of the body in minutes. The intersection of the circles then "fixes" our position.

Actually, this is all there is to celestial navigation. Unfortunately, however, practical navigation problems cannot be solved by this simple graphical method. The distances involved, and the physical limitations of chart size and space upon which to lay them, make such a graphical solution impractical. In practice, navigation problems are solved, using this identical theory, but by approaching the solution from a different direction.

We'll get into this practical method soon, but first let's talk about two "sights" that have guided sailors over the oceans of the world for centuries. These methods are routinely employed by practical navigators throughout the world, and are included in the survival manuals of the military.

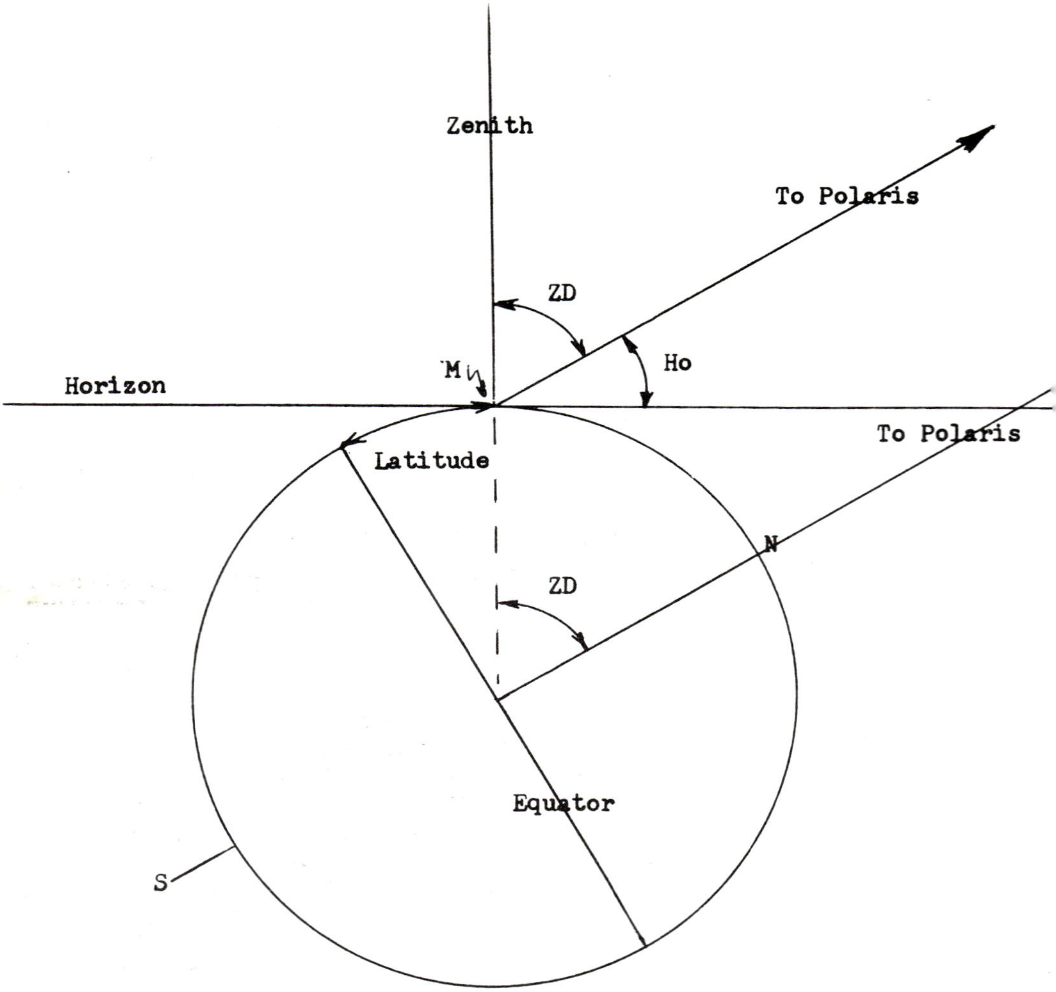

ZD + Ho = 90°
ZD + Latitude = 90°

Therefore

Ho = Latitude of observer

FIGURE 19

THEORY OF A POLARIS SIGHT 10

Perhaps the oldest navigational star, and the most important, is the Pole Star, or Polaris. This star, known to all who live in the Northern Hemisphere, has a declination of 89°08′.5 North. Its GP, therefore, is less than one degree from the North Pole. Assume, for the moment, that Polaris is exactly over the North Pole. To a person standing there, Polaris is directly overhead. To a person standing at the equator, Polaris is on the horizon. To a person standing somewhere in between, the altitude of Polaris is the latitude of the observer (FIG. 19). Therefore, for a navigator to know his latitude, all he has to do is measure the altitude of Polaris. Practically, a small correction must be made due to the small circle that Polaris makes around the North Celestial Pole. This correction is listed in the almanac as a function of the LHA of Aries at the time of the observation. The correction varies from 0 to plus-or-minus 51 minutes.

The detail of working out an actual Polaris sight will be presented in Chapter 17.

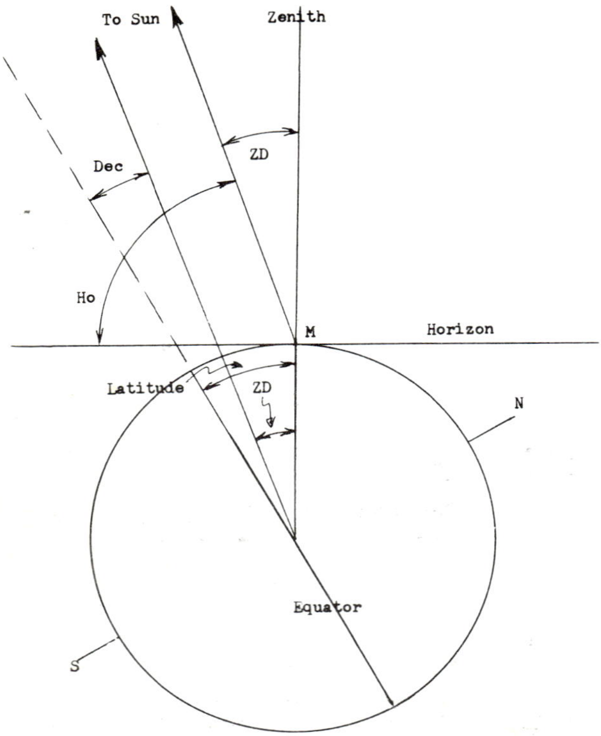

ZD + Declination (N) = Latitude of observer

FIGURE 20

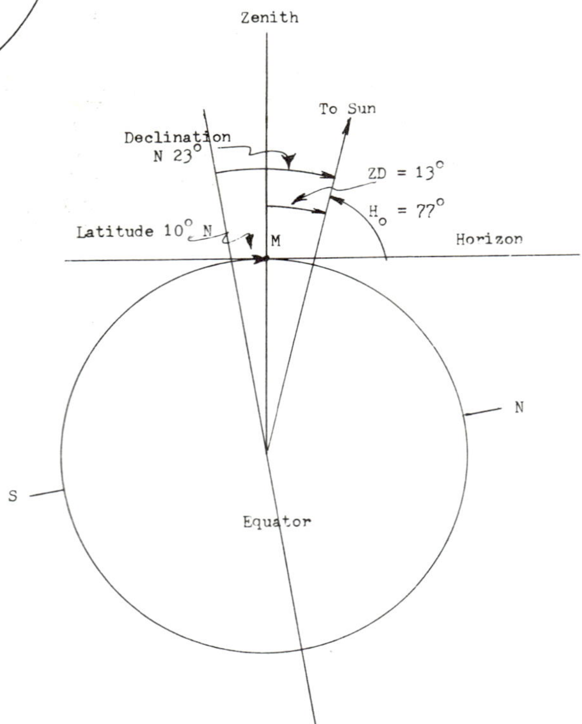

FIGURE 21

Declination - ZD = Latitude of observer

THEORY OF A NOON SIGHT 11

As we learned earlier, an observer on earth sees the sun rise in the east, move upward to its highest point on the meridian at noon, and then start moving down toward the west. If the altitude of the sun is measured at its highest position, and the declination of the sun is known, then the latitude of the observer can be determined by simple arithmetic, without reference to navigation tables, and without having to know the exact time. Figure 20 illustrates that Zenith Distance of the sun, on the meridian, **plus** the declination of the sun, equals the latitude of the observer. (During the half of the year that the sun has a south declination, Latitude equals Zenith Distance **minus** Declination.)

To obtain your latitude, start taking altitude sights of the sun as it approaches the meridian. When it finally reaches the highest position and starts to go down, note the highest sextant reading, subtract this reading from 90°, to obtain the Zenith Distance, add (or subtract) the sun's declination, and you have the latitude directly.

In a special situation, such as when the sun's declination is greater than the latitude of the observer, the sun is north of the observer at noon. (Northern Hemisphere observation.) In this case, Declination minus Zenith Distance equals the Latitude of the observer. Figure 21 shows this relationship. In this example Declination equals N 23°, Zenith Distance equals 13°, and the Observer's Latitude equals 10° North. (23° - 13° = 10° N. Lat.)

Whenever any special situation arises, or you are in doubt about a solution, draw a simple diagram such as Figure 21. The answer will then be obvious.

We'll work up an actual noon sight in Chapter 16.

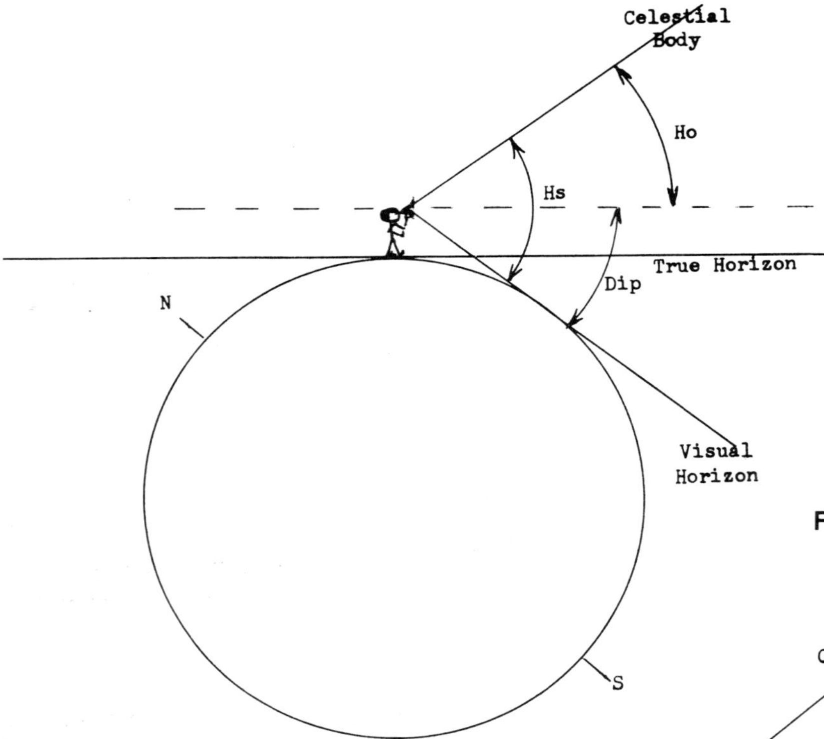

FIGURE 22

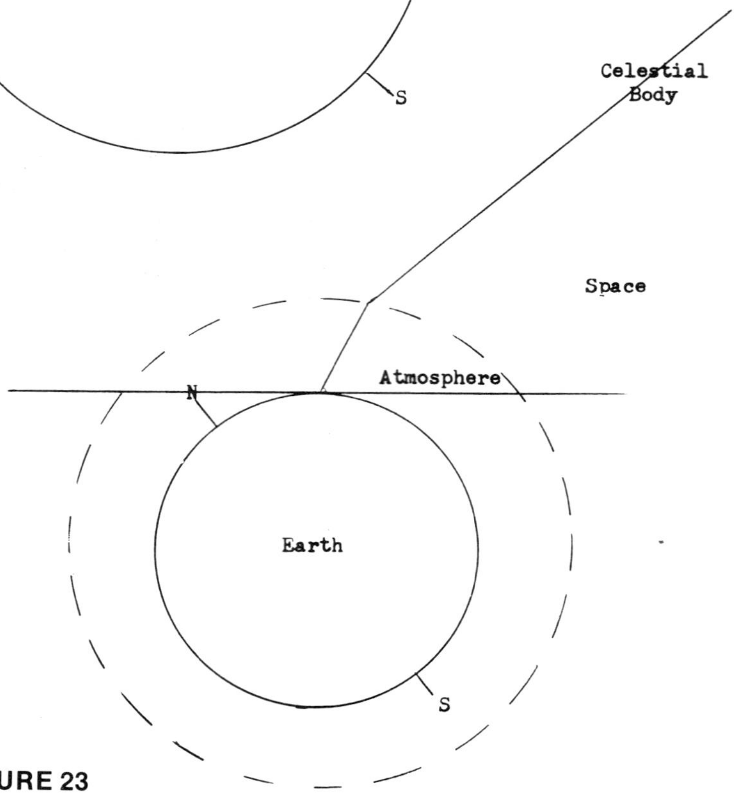

FIGURE 23

CORRECTIONS TO SEXTANT READINGS 12

All celestial navigation problems start by first measuring the altitude of a celestial body above the horizon. This angle is called Hs, meaning the angular height (H) of the body as measured by the sextant (s).

The usual marine sextant is an instrument consisting of mirrors mounted on a frame that can be adjusted to measure angles. These angles are then read from a calibrated scale on the instrument. When one looks through the sextant, the celestial body and the horizon are in the field of view. By adjusting the sextant, the celestial body is made to appear in exact contact with the horizon. In the case of the sun and moon, either the upper limb or the lower limb of the body is brought into contact with the horizon.

In the case of sights made on planets or stars, since they have no apparent diameter, their spot of light is brought down and aligned with the horizon.

However, before the sextant height Hs can be used, certain corrections must be made to the reading. These corrections are small, generally only a few minutes of arc. We will consider them in the order in which they are usually applied.

First - Sextant Correction

The sextant itself may have some small optical or mechanical error in construction. If a correction is needed, it is noted as a "Sextant Correction" (SC) by the manufacturer. This correction could be positive or negative to be added to or subtracted from Hs.

Second - Index Correction

The "Index Correction" (or zero reading) of the sextant must be checked before making an observation. Set the sextant for "zero altitude" on the instrument's scale. Then view the horizon. If the two sections of the horizon seen through the sextant do not coincide exactly, the sextant is adjusted until they do. If the sextant then reads below zero, the amount below zero must be added to a later observed Hs. If the sextant reads above zero when the horizons coincide, then the amount above zero must be subtracted from a later observed Hs. The "Index Correction" (IC) can be either positive or negative, depending on the particular sextant and its adjustment.

Third - Dip

As noted previously, the altitude of a celestial body is measured as an angle above the horizon of the observer. This is an angle measured from the plane of the observer or a plane tangent to the surface of the sea at the observer's position. If the observer is above the surface of the sea, as is always the case because of his height or elevated position on deck, he actually looks over the curvature of the earth. The apparent horizon "dips" below the true horizon (FIG. 22). Thus any altitude measured with a marine sextant actually is greater than if it were measured from the surface. A "dip correction" is therefore necessary. The "dip" is always negative and depends on the height of the observer. On a small boat, with observations being made between six and twelve feet above the water, the correction is -3′ of arc. "Dip" is tabulated on the back cover of the Air Almanac. (See Table I, Page 77.)

Fourth - Refraction

Light from a celestial body travels from a body in space to the observer on earth. The light travels through the vacuum of space and then through the atmosphere of the earth to the observer's eye. Whenever light passes at an angle through matter of different density, it does not travel in a straight line. The path of the light bends, or is subject to refraction. This is the same physical phenomenon that is observed in the apparent bending of a stick of wood when it is put into water at an angle. Refraction causes the light from a celestial body to bend down toward the observer, and always **increases** the apparent altitude of the body (FIG. 23). Refraction diminishes with observations approaching the zenith, and increases to a maximum at the horizon.

The "Refraction Correction" is always negative, and is subtracted from Hs. The "Refraction Correction" is listed on the inside back cover of the Air Almanac. (See Table II, Page 77.)

Fifth - Semi-Diameter

Astronomical tables and almanacs always tabulate altitudes of celestial bodies measured "center to center," from the center of the earth to the center of the body. Two celestial bodies, the sun and the moon, as viewed from the earth, are so large that they are not points of light but large luminous discs in the sky. They have a measurable diameter seen through a sextant. In making observations of either the sun or the moon, as noted

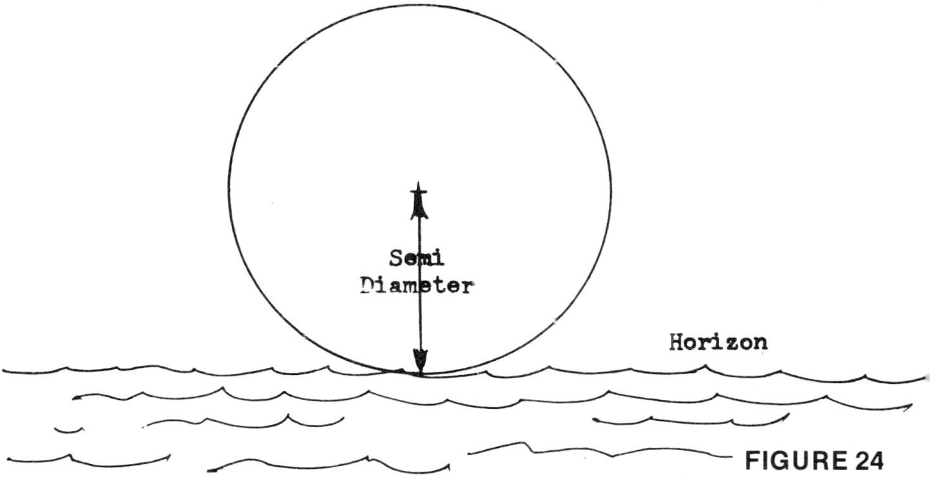

FIGURE 24

previously, the sextant is adjusted so that either the upper or lower limb (rim) of the body is brought into contact with the horizon (FIG. 24). Thus, to correct the altitude observed with the sextant in order to bring the observatin to the middle of the body, a "semi-diameter" correction must be applied. In the case of a "lower limb" observation, the correction must be **added** to Hs; in the case of an "upper limb" observation, the correction must be **subtracted** from Hs. The semi-diameter correction (SD) for both the sun and the moon is listed on each daily page in the Air Almanac. (See Table III, Page 78.)

Sixth - Parallax

As noted earlier, the positions of celestial bodies relative to the earth are tabulated in tables and almanacs "center to center." Obviously, an observer on the surface of the earth is not making observations from the center of the earth. However, astronomical distances are so great that with the exception of observations made on the moon, which is only about a quarter of a million miles from us, it makes no difference whether an observer is at the poles, or at the equator, or some place in between. Not so, however, with observations of the moon, which is so close to us. When an observer moves about the earth, with a diameter of about 8,000 miles, only a quarter of a million miles from the moon, he changes his angle to the moon appreciably. This change in angle with position is called **Parallax in Altitude**.

To demonstrate the effect of parallax, hold your finger out at arm's length and look at a distant object. Look first with one eye and then with the other, simulating a sight from one location on the earth and then from another. The change in sighting angle, or parallax, is very apparent.

The moon's Parallax in Altitude, or "P in A," is listed on each daily page in the Air Almanac. The correction is always positive. (See Table III, Page 78.)

When all of the applicable corrections to the sextant altitude Hs are made, the **Observed Altitude Ho** has been obtained.

TIME OF SIGHT 13

Correct time is essential to celestial navigation. The GP of a celestial body goes completely around the earth in 24 hours, or it moves over the surface of the earth at a rate of 900 knots when at the equator. This breaks down to the GP moving as much as one nautical mile every four seconds. So the accuracy of measuring your position depends on the accuracy of the time when you make your sight.

Fortunately, the modern small transistor radio is able to receive "time-ticks" that are broadcast continuously giving GMT. Time is no longer a problem for navigators. Time signals broadcast by the National Bureau of Standards from WWV, Fort Collins, Colorado, are heard on 2.5, 5, 10, 15, 20, and 25 mhz. WWVH, Kekaha, Kauai, Hawaii, broadcasts on 2.5, 5, 10, 15, and 20 mhz. The particular station and frequency that provides the best reception depends on the time of day, location, and atmospheric conditions. A good signal from some station can be obtained at any time.[2]

With a portable shortwave radio and a common stop-watch, extremely accurate time can be obtained. The stop-watch is started at the exact "tick" of an announced minute. The watch is stopped at the exact time a celestial measurement is made. The time on the stop-watch, added to the time of the "tick," gives the GMT of the observation.

The average navigator can start and stop a stop-watch without any appreciable error. Since the time broadcast by the Bureau of Standards radio stations is announced in GMT,[3] no conversion from local time is required. The GMT of the observation is obtained directly.

2. Most Maritime nations broadcast time signals. "Radio Navigational Aids, H.O. Pubs. Nos. 117-A and 117-B" list them all. The U.S. Navy transmits time signals over a number of Atlantic and Pacific stations; the British Broadcasting Corporation directs time signals to North America, Africa, Asia, the Far East, NE & NW Pacific, Australia, and the SW Pacific; Argentina, Australia, Canada, Japan, the Republic of South Africa, and Switzerland provide time signal transmissions heard over a very wide area.

3. Effective January 1, 1974, as a result of an international conference in which the United States participated, WWV and WWVH now designate the time as "Coordinated Universal Time, UTC," in place of the previously described "Greenwich Mean Time." The name only is changed. The new designation may again be changed in the future.

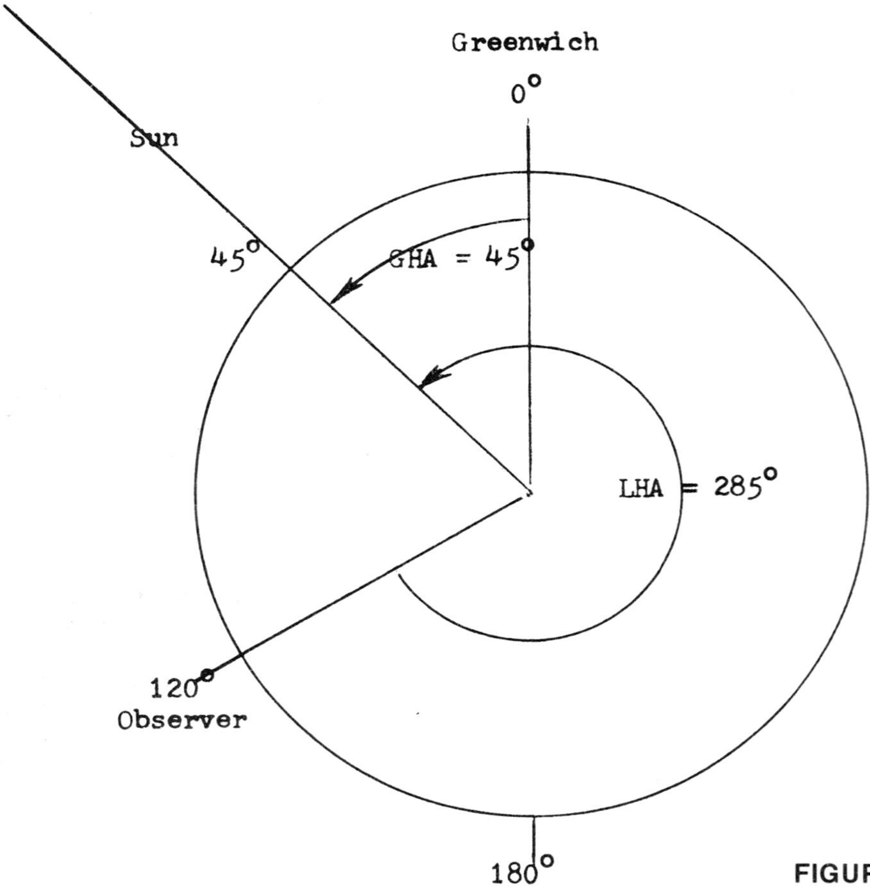

FIGURE 25

TIME DIAGRAMS 14

Whenever a navigator works up a sight, it is good practice to draw a time diagram of the problem. This practice eliminates errors and simplifies work.

A time diagram is a circle representing the earth as viewed from the South Pole looking north through the earth's axis (FIG. 25). At the top of the circle is the Greenwich meridian 0°. The Greenwich Hour Angle (GHA) of the body is indicated at the time of the sight. Remember that GMT starts at the 180th meridian, or just halfway around the world from Greenwich, in the Pacific.

Suppose, for example, that the time is 1500 GMT as broadcast over WWV. This means that the sun is 15 hours from the 180th meridian, or 3 hours west of Greenwich. Since 1 hour of t me represents 15°, the sun is 45° west of Greenwich. This angle, 45°, is known as the Greenwich Hour Angle (GHA) of the sun.

The Air Almanac lists the GHA of the sun, Aries, the navigational planets, and the moon, for every second of GMT of every day.

After the GHA has been obtained from the Air Almanac and drawn on the time diagram, the Local Hour Angle (LHA) is determined. The LHA is the angular distance from the observer's meridian to the body, measured **westerly.**

Suppose, in this example, that the observer is off the coast of Southern California at a dead reckoning position of 120° West Longitude. This position is indicated on the time diagram Figure 25. The LHA is the angle between 120°, around to the sun's position at 45° (measured **westerly**). LHA equals 285°.

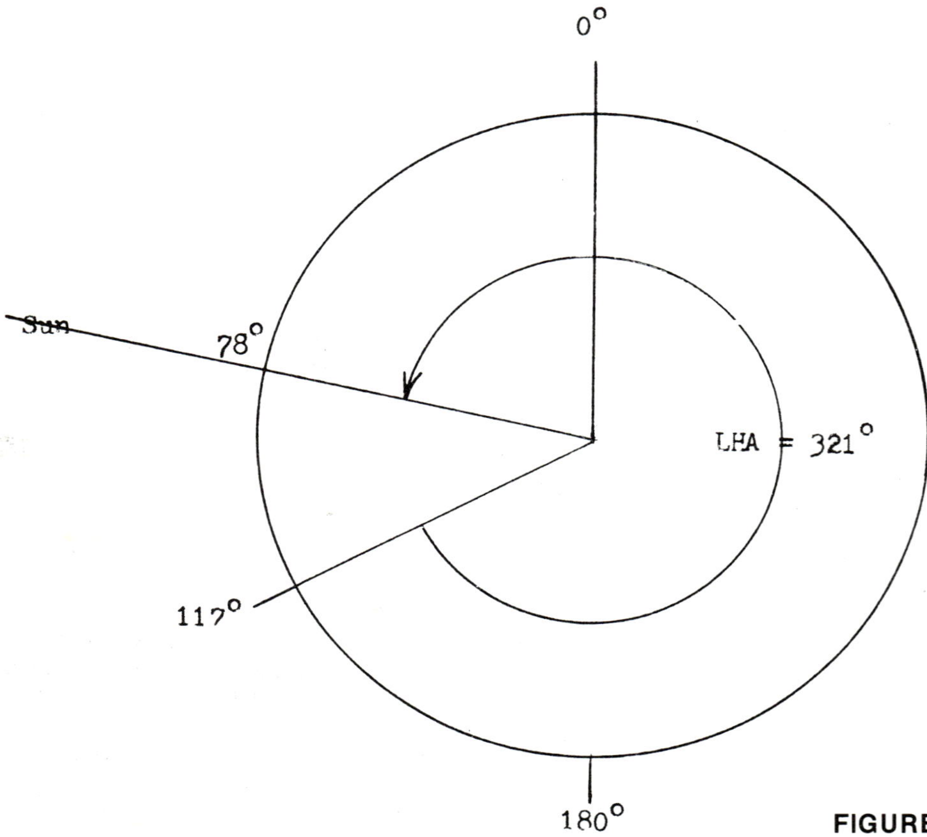

FIGURE 26

NOTE: In the following chapters, reference is made to Tables I - VIII. These tables are found on pages 77 - 83.

PRACTICAL CELESTIAL NAVIGATION 15

Back in Chapter 9 we learned that by plotting "circles of position" on a chart we could determine our position at sea. We said, however, that in actual practice there was a better way of doing it. So let's look into that better way now.

Suppose we are sailing off the Southern California coast. The date is October 1, 1973, and it is mid-morning. Our dead reckoning position is 117°30′ West Longitude, and 32°45′ North Latitude. The time is about 10 AM PDT, or 1700 GMT.

The sun is shining brightly and we decide to obtain a "line of position." Tuning our radio to WWV, we listen to the "time-ticks" being broadcast. We take out our sextant, check the Index Correction as (-2′), and hang the stop-watch around our neck. We start the stop-watch when WWV says it is 1702 GMT.

In making the sextant observation, the navigator faces the sun. He takes up a position in the cockpit, or in the companionway, or leaning against the mast or stays, in such a manner that he can brace himself comfortably while keeping both of his hands free to hold and adjust the sextant. He must have a clear view of the sun and the horizon without masts, stays, lines, stanchions, etc., being in the field of view. Some navigators attach a neck lanyard or wrist strap to the sextant to prevent dropping the instrument in case of a sudden lurch of the boat or other accident. The navigator then adjusts the upper and lower filters of the sextant so that the sun is a good clear dull ball, and the horizon is clean and sharp.

We have taken up a position with our eye about eight feet above the water line. Holding the sextant as vertical as possible, the instrument is adjusted until the lower limb of the sun is in exact contact with the horizon. The sextant is then moved slowly back and forth, from left to right, causing the image of the sun to move like a pendulum across the horizon. This is called "swinging the arc" or "rocking the sextant," and insures that the sextant is being held vertical. The sextant is further adjusted until the sun is in exact contact with the horizon when the sun is at its lowest "swing." The stop watch is clicked off. It reads 1 minute 20 seconds at the time of the sight.

The sextant reads: Hs 37°57′

(In working up our sights we'll round off all angles read from the sextant scale, and all figures taken from the Air Almanac, to the nearest minute of arc. The limiting accuracy for small boat observations, as noted in the Preface, is thus one minute of arc, or one nautical mile.)

So far we have the following information:

Body	Sun
Date	October 1, 1973
DR position	117°30°W - 32°45′
Time	17:03:20 GMT
Hs	37°57′

Let us now make the corrections to Hs that were discussed in Chapter 12. You will remember that they were: Sextant Correction; Index Correction; Dip; Refraction; Semi-Diameter; and Parallax in Altitude.

We will assume that the sextant is well made and has no SC. The IC we checked as (-2′). The Dip, as listed on the back cover of the Air Almanac for a height of 8 feet, is (-3′), (Table I). Refraction, found on the inside back cover of the Air Almanac (Table II) for **"0 height above sea level;** (the other columns are used only in airplane observations at various airplane altitudes), is (-1′). Semi-Diameter is listed at the bottom of each daily page in the Air Almanac for the Sun as (+16′), (Table III). We do not correct for P in A since it applies only to moon sights.

We now have the following information for obtaining the observed altitude Ho.

Hs		37°57′
	SC	0′
	IC	-2′
	Dip	-3′
	Refr	-1′
	SD	+16′
Ho		38°07′

The Air Almanac tabulates the GHA of the Sun, Aries, the Navigational Planets, and the Moon, for every 10 minutes of GMT, for each day in the year. Our observation was made at 17:03:20 GMT.

The GHA of the sun is tabulated in the Air Almanac for 1700 GMT as 77°36′ (Table III). Declination of the sun at that time is South 3°19′ (rounded off).

But since our observation was made at 17:03:20 GMT, or 3 minutes 20 seconds later than the time tabulated in the daily pages, we must interpolate to obtain the value of GHA for 17:03:20 GMT.

There are two identical "Interpolation of GHA" tables in the Air Almanac (Table IV). One is on the inside front cover, and the other is on the "flap" just behind the daily pages. It is customary for navigators to tear out each daily page in the Air Almanac when that day is over. Thus when using the AM page for a particular date, the Interpolation Table on the inside front cover is in view. When using the PM page, the "flap" is flipped out and, in our case, is placed over the AM page for October 2, 1973. It is thus next to the PM page of October 1, 1973, which is the page we are using.

Referring to the "flap" we note the following:

GHA for 1700 GMT (from the daily page)	77°36′
Increment for 03:20 (from table on flap)	0°50′
GHA of Sun for 17:03:20 GMT	78°26′

The declination change for this additional time increment is so small that no correction is needed.

We now have all the information we need from the Air Almanac, so we put it aside.

We said in Chapter 14 that a "time diagram" should always be drawn (FIG. 26). Round off the figures to make it easy to do the arithmetic. We just want to "see" the problem visually, and we won't actually use these specific figures in solving the problem. We are at 117° W. Long. The GHA of the sun is 78°. The LHA of the sun is 321°. The sun is east of our position by 39° (117° -78°). The "time diagram" makes this clear.

Before we pick up HO249 to finish working up the sight, a little more theory is in order for us to better understand what we are going to do.

We have determined the GP of the sun from the Air Almanac. (78°26′ W. Long. :3°19′ S. Lat.). We know the sun's altitude from our position at sea. If we had a large enough chart, we could now draw a circle on the chart with the center at the GP of the sun, and a radius of 3113 nautical miles. (Zenith Distance is 90° minus 38°07′ or 51°53′ or 3113′ which corresponds to 3113 nautical miles.) This would give us our "circle of position" on the chart. But this is not practical, as we said in Chapter 9.

To get around this, we use "Sight Reduction Tables." We said earlier that astronomers can calculate the altitude of any heavenly body, from any spot on earth, for any instant of time. They can also calculate the direction the body is from the assumed position. In other words, using spherical trigonometry, they solve the navigational triangle and tabulate the results.

In the following hypothetical case, suppose we determine by dead reckoning that we are at 117°00′ W. Long. - 33°00′ N. Lat. We observe the sun. Ho is 52°20′. We look in a "Sight Reduction Table" and find that the altitude of the sun, **as calculated for our DR location** at this specific time, is tabulated as Hc = 52°20′, the same value! We know then that we are on a "circle of position" that passes exactly through our DR position.

Suppose, however, that when we looked in the table, it said that the calculated Hc is 52°30′. What would that tell us? It would simply mean that we were 10 nautical miles further **AWAY** from the sun than we thought. Our actually observed Ho is 52°20′. We will have to move 10 nautical miles closer to the sun to obtain a sextant reading Ho of 52°30′. Hc minus Ho

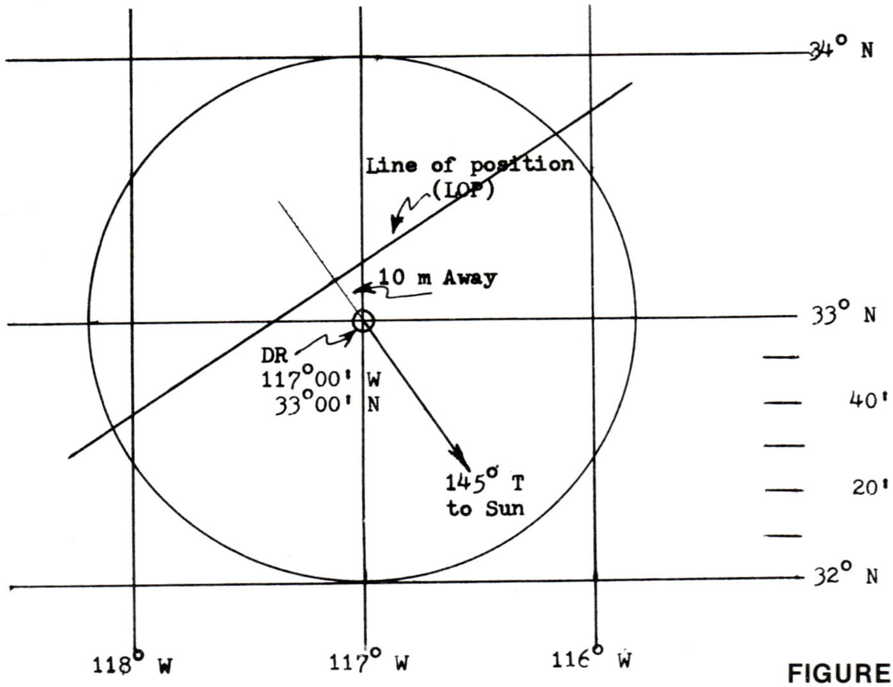

FIGURE 27

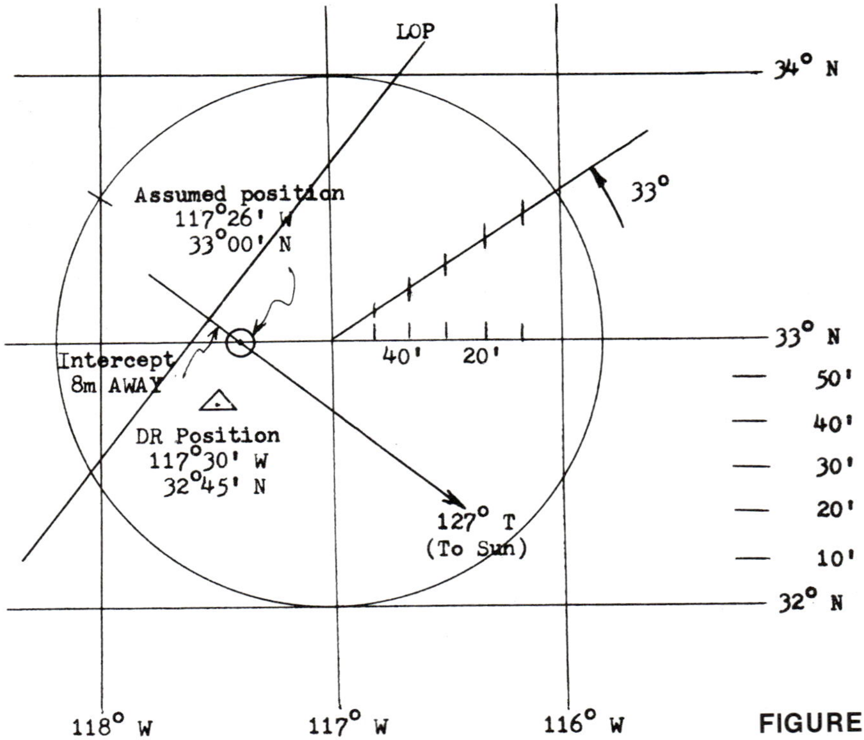

FIGURE 28

equals 10 minutes. 10 minutes equals 10 nautical miles. The calculated altitude is **correct** for our DR position, so the DR position has to be incorrect. We are actually 10 nautical miles **"AWAY"** from the DR position.

The tables tell us something else. They tell us the azimuth from our assumed position to the sun. From our assumed position we can draw a line on the chart in the direction of the sun. Let us assume, in this hypothetical case, that the azimuth is 145°T.

Let us now refer to Figure 27. This is a "plotting chart" indicating a portion of the sea from 116° to 118° West Longitude, and from 32° to 34° North Latitude. Our DR position of 117°00′ W. and 33°00′ N. is indicated. The angle, or azimuth, from this DR position to the sun, is shown.

The altitude of the sun, as we observed it from the DR position, was Ho = 52°20′. When we looked in the tables, however, the altitude should have been Hc = 52°30′. We were, therefore, not at the DR position when we took the sight. Our correct position was 10 nautical miles **AWAY** from the DR position. (Hc - Ho = 10)

To correct for this, we measure a distance of 10 nautical miles **AWAY** from the DR position along the azimuth line. We then draw the LOP at right angles to this line. We are somewhere on this LOP.

This is how practical celestial navigation problems are solved.

(Note: In any practical celestial navigation problem we are a long distance away from the GP of the sun. The radius of our "circle of position" is therefore very long. The short length of the circumference of the "circle of position" we draw on the plotting chart can be considered to be a straight line. Thus the "LOP" is drawn as a straight line as indicated above.)

For practical considerations, Sight Reduction Tables are limited in the information they can contain. To tabulate altitudes and azimuths from every spot on earth (every second of latitude and longitude), would require hundreds of volumes of tables. They could not be carried on a small boat.

HO249 tables are the most compact of the Sight Reduction Tables. They list Declinations in whole degrees from 0° to 29°; with Assumed Latitudes in whole degrees from 0° to 39° North and South in Volume II, and Assumed Latitudes 40° to 89° in Volume III. LHA is tabulated in whole degrees. (For latitude 70° and above the LHA interval is two degrees.)

Using mathematical language, the "Entering Arguments" needed to use HO249 are : Latitude, whole degrees; LHA, whole degrees; Declination, whole degrees. "Argument" is merely the name or the heading of the page, column, and line you use in a table.

While we are discussing language, Declination **SAME** NAME AS LATITUDE, means the declination of the body is North if your assumed position is in North Latitude, and the declination of the body is South if your assumed position is in South Latitude. Declination **CONTRARY** NAME AS LATITUDE, means Declination and Assumed Latitude are not both the same: one is North and the other is South.

(Volume I, HO249 is used with selected stars, and we'll talk about this volume later.)

Now let's get back to our original problem. This is the information we have so far:

WORK SHEET

Body ______________ Date ______________ DR (Long. ______________ / (Lat. ______________

Time of sight ______________________ GMT

GHA(*) _______ at _________ GMT _________

(*) GHA of Body, or GHA of Aries for star sights.

Increment for _________ _________

GHA(*) _______ _________

Declination ______________

Hs ___________

SC _______

IC _______

Dip _______

Refr _______

SD _______

P in A _______

Ho ___________

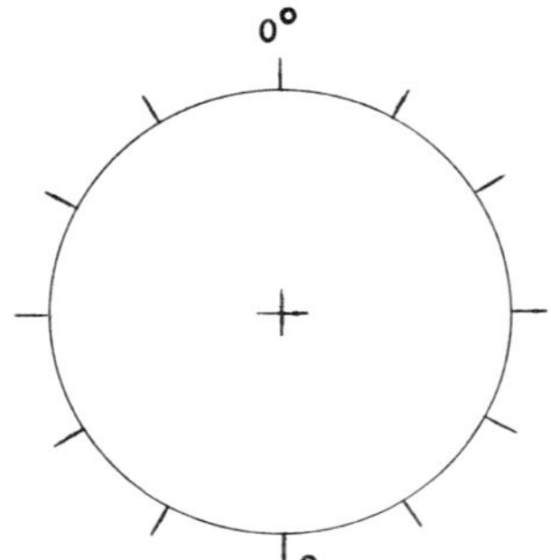

TIME DIAGRAM

GHA(*) _______ _______ + (360°) _________

Assumed Long. _________

LHA _________

Assumed Lat. _________

Declination Argument _________ (Same) (Contrary)

HO249 Hc ___________ d ___________ Z ___________

d correction _______

Hc ___________ Zn ___________

Hc - Ho ___________ (Away) + (Toward) -

NOON SIGHT

Latitude = ZD ± Declination

= 89° 60'

(-) _______ (Ho)

= _______ (ZD)

(±) _______ Declination

Latitude = ___________

POLARIS SIGHT

Ho ___________

Q for LHA ♈ _____ ___________

Latitude ___________

FIGURE 29

Body	Sun
Date	October 1, 1973
DR Position	117°30′W 32°45′N
Time	17:03:20 GMT
Ho	38°07′
Declination of Sun	S3°19′
GHA of Sun	78°26′

We now want to enter Volume II of HO249, so we must determine our "Entering Arguments." Latitude must be a whole degree, the sun's Declination must be a whole degree, and LHA must be a whole degree. We will chose:

Latitude 33°N (close to our DR position)

Declination S 3° (next whole degree less than actual value)

LHA (from our time diagram Figure 26) is the angular distance **westerly** from our position, to the sun's meridian. So with the GHA = 78°26′, or 78°26′ + 360° = 438°26′ (add or subtract 360° anytime you wish to make the arithmetic easy), we subtract an assumed longitude close to our DR position so that the LHA will come out to a whole degree. 438°26′ - 117°26′ = 321°.

Our **Assumed Position** is thus: 33°N. Lat. - 117°26′ W. Long.

The "Entering Arguments" for HO249 Volume II are:

Latitude	33° North
Declination	3° South
LHA	321°

We open HO249 Volume II to the page "Lat 33°" and use the page (200) that is marked "DECLINATION (0° - 14°) **CONTRARY** NAME AS LATITUDE" (Table V). Going down the column for 3° Declination, to the line LHA 321°, we find the following information tabulated:

Hc	38°29′
d	-44
Z	127°

We are attempting to obtain an Hc to compare with our Ho. The Hc we have now obtained is for a Declination of 3°. Our actual Declination is 3°19′. We are going to interpolate for the extra 19′. This is where the "d" comes in.

Table 5, in the back of HO249, page 242, gives you the answer (Table VI). Look across the top of the Table to (d) = 44. Then go down that column to (′) = 19. The answer is listed as 14′. Subtract this 14′ (since d is -) to get the corrected Hc.

Hc = 38°29′ - 14′ = 38°15′

(It is best to use Table 5 in HO249, but you can figure the "d" correction if you wish. If we had used the column "Declination 4°," instead of "Declination 3°," in HO 249, Hc would be 37°45′. This is 44′ less than the "Declination 3°" figure of 38°29′. The actual declination we are concerned with is 3°19′, or 19/60 of the difference between 3° and 4°. 19/60 of 44′ is 14′).

We compare Observed Height with the Computed Height.

Hc: 38°15′ - Ho: 38°07′ = 8′

The "Intercept" is 8 nautical miles

WORK SHEET

Body SUN Date 1 OCT 73

DR (Long. 117°30' W
DR (Lat. 32°45' N

Time of sight 17:03:20 GMT

GHA(*) SUN at 1700 GMT 77°36'

(*) GHA of Body, or
GHA of Aries for star sight

Increment for 03:20 50'

GHA(*) SUN 78°26'

Declination S 3°19'

Hs 37°57'

SC ~~0~~

IC -02'

Dip -03'

Refr -01'

SD +16'

P in A —

Ho 38°07'

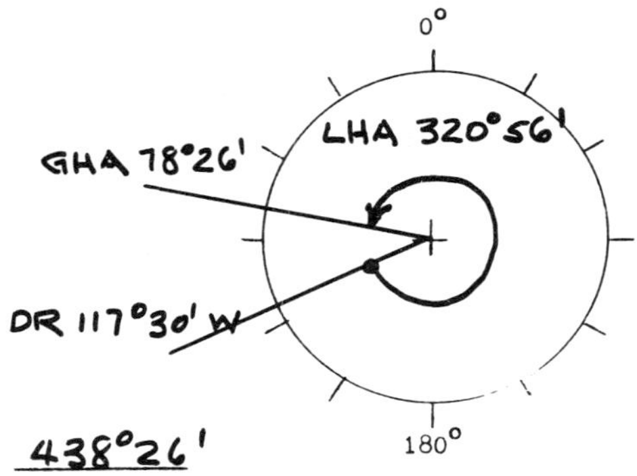

TIME DIAGRAM

GHA(*) SUN 78°26' + (360°) 438°26'

Assumed Longitude 117°26'

LHA 321°

Assumed Latitude 33° N

Declination Argument 3° ~~(Same)~~ (Contrary)

HO249 Hc 38°29' d -44 Z 127°

d correction -14'

Hc 38°15' Zn 127°T

Hc - Ho +8' (Away) + ~~(Toward)~~ -

NOON SIGHT

Latitude = ZD ± Declination

89° 60'

(-Ho) ______

(=ZD) ______

(±Dec) ______

Latitude = ______

POLARIS SIGHT

Ho ______

Q for LHA ♈ ______ ______

Latitude ______

FIGURE 30

Since the observed height is **less** than the computed height, we move our position 8 nautical miles **AWAY** from the sun from the assumed position of 33°N - 117°26′W.

"Z," the azimuth angle of 127°, is the angle from our position that the sun makes from the nearest pole. "Zn," which is the azimuth measured from true north, is needed when you plot a position on a chart. In our problem, we know that the sun was east and south of us, so there is no confusion. But on the top and bottom left hand side of each page in HO249 (Table V), is a rule that will always give you the right answer for Zn. In our case we use the top line which reads: **"N. Lat. LHA greater than 180° ... ZN = Z."** So the bearing from our assumed position to the sun is 127°T. (In all celestial navigation problems, TRUE headings and TRUE bearings must always be used.)

Now let's put the problem on the chart.

If you don't want to do all your scribbling on your regular navigation chart, which can get messy and confusing if you work in one area several times, it might be well to use plotting charts and then transfer your fix to your regular chart. If you don't have a supply of plotting charts, or a "navigation notebook," you can draw one in a few seconds (Fig. 28).

Draw a circle with a radius of six half inches, or six centimeters, or six units of some scale you have. You'll see in a minute that by using the radius of the circle as one whole degree of latitude, each unit of the scale used is 10′ of latitude, or 10 nautical miles. This makes plotting on the chart easier.

The longitude scale is not a full radius, since as latitude increases the longitude distance decreases. (The orange segment effect.)

Again, in our example, the assumed latitude was 33°N. Make a mark on the circumference of the circle 33° above the horizontal diameter. Draw vertical lines of longitude through these 33° marks. Label the three latitude lines, and the three longitude lines, and plot your data. That's all there is to making your own plotting chart, and plotting your line of position.

It has taken a long time and many pages to explain this navigation problem. It is harder to describe the process than it is to do it. In practice the procedure is short, and easy. A single "Work Sheet," similar to Figure 29 is all that is needed for the data. Perhaps you might want to mimeograph or print your own to use. The sheet is quickly filled in and then from this single page your line of position is plotted as in Figure 28. The whole process takes but a few minutes. You may now want to re-read this chapter and compare the data as it is developed with a completed "Work Sheet" (Fig. 30).[4]

4. A simple notation on the "Work Sheet" eliminates the necessity of "thinking out" whether the intercept is "Away" or "Toward" the sun. If Hc is greater than Ho, then Hc - Ho is positive (+) and the intercept is "Away." If Hc is less than Ho, then Hc - Ho is negative (-) and the intercept is 'Toward." This is noted on the "Work Sheet," Figure 29.

WORK SHEET

Body SUN Date 1 OCT 73

Time of sight 1930 - 1940 GMT

GHA(*) ______ at NOON GMT ______

Increment for ______ ______

GHA(*) SUN 117°30'

DR (Long. 117°30' W
(Lat. 32°32' N

(*) GHA of Body, or
GHA of Aries for star sight

Declination S 3°22'

Hs 53°54'

SC —0—

IC -02'

Dip -03'

Refr -01'

SD +16'

P in A —

Ho 54°04'

0°

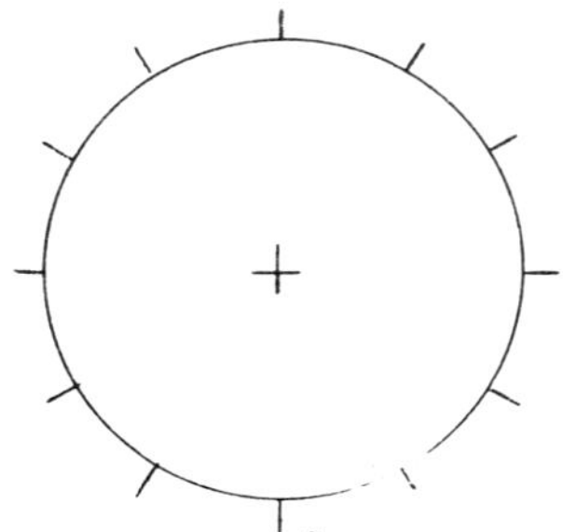

180°

TIME DIAGRAM

GHA(*) ______ ______ + (360°) ______

Assumed Longitude ______

LHA ______

Assumed Latitude ______

Declination Argument ______ (Same) (Contrary)

HO249 Hc ______ d ______ Z ______

d correction ______

Hc ______ Zn ______

Hc - Ho ______ (Away) + (Toward) -

NOON SIGHT

Latitude = ZD ± Declination

89° 60'

(-Ho) 54°04'

(=ZD) 35°56'

(±Dec) - 3°22'

Latitude = 32°34' NORTH

POLARIS SIGHT

Ho ______

Q for LHA ♈ ______ ______

Latitude ______

FIGURE 31

THE NOON SIGHT 16

In Chapter 11 we talked about the "Noon Sight" as one of the old navigational methods employed by sailors as far back as the Fifteenth Century. Let's illustrate this method as it is practiced today.

The date is October 1, 1973, and the DR position is 117°30′ W - 32°32′ N, and it is near noon. First we must determine when it will be noon **local time**, or when the sun will cross the observer's meridian. This will occur when the GHA of the sun is the same as the longitude of the observer. We open the Air Almanac for October 1, 1973 (Table III) and run down the column for the GHA of the sun. We find that at 1930 GMT, the GHA of the sun is 115°06′.2. At 1940 GMT, the GHA of the sun is 117°36′.3. Thus, the sun will cross the meridian at 117°30′ W sometime between these two times. Since our watch is set on PDT, the meridian crossing will occur between 12:30 PM and 12:40 PM PDT. (GMT minus 7 hours)

Exact time is not important for this sight. We take out our sextant, check the Index Correction, and start observing the altitude of the sun before meridian crossing about 12:30 PM PDT. When the highest altitude of the sun is observed, and it starts to go down, we note the highest reading obtained. Hs = 53°54′.

We work up the sight as follows:

Hs	53°54′
IC	-02′
Dip	-03′
Refraction	-01′
SD	+16′
Ho	54°04′

Declination of sun (from Air Almanac) S 3°22′ (rounded off)

Latitude = Zenith Distance - Declination
= (90° - 54°04′) - 3°22′
= 35°56′ - 3°22′

Latitude = 32°34′ North

Or we are 2 nautical miles north of our DR position.

The data is tabulated on the "Work Sheet" using appropriate portions of the page (FIG. 31).

WORK SHEET

Body POLARIS Date 2 OCT 73

DR (Long. 117°14' W (Lat. 32°24' N

Time of sight 0200 GMT

GHA(*) ♈ at 0200 GMT 40°39'

(*) GHA of Body, or GHA of Aries for star sight

Increment for —

GHA(*) ARIES 40°39'

Declination ______

Hs 32°14'

SC —0—

IC -02'

Dip -03'

Refr - 02'

SD —

P in A —

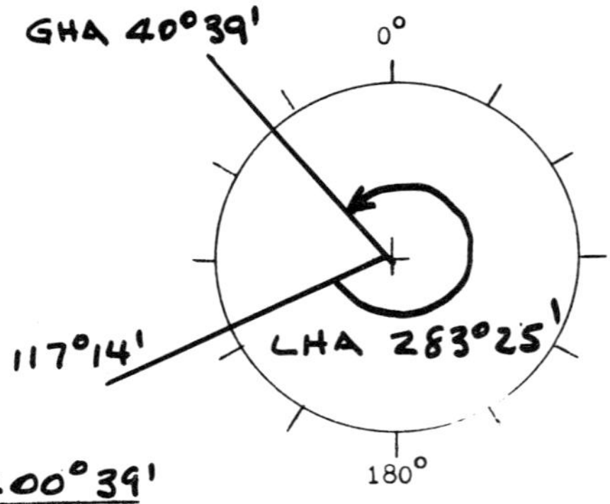

TIME DIAGRAM

Ho 32°07'

GHA(*) ARIES 40°39' + (360°) 400°39'

~~Assumed~~ Longitude 117°14'

~~LHA~~ 283°25'

Assumed Latitude ______

Declination Argument ______ (Same) (Contrary)

HO249 Hc ______ d ______ Z ______

d correction ______

Hc ______ Zn ______

Hc - Ho ______ (Away) + (Toward) -

NOON SIGHT

Latitude = ZD ± Declination

89° 60'

(-Ho) ______

(=ZD) ______

(±Dec) ______

Latitude = ______

POLARIS SIGHT

Ho 32°07'

Q for LHA ♈ 283°25' + 16'

Latitude 32°23' N

FIGURE 33

LATITUDE BY POLARIS 17

We sail south in the late afternoon and "drop the hook" in the lee of South Coronado Island. We know where we are, but just for the practice we'll take a sight on Polaris and work out our latitude by this method. We are at 117°14′ W - 32°24′ N.

It is dusk, as we bring Polaris into tangency with the horizon at 7:00 PM PDT. The sextant reads 32°14′.

Hs		32°14′
	IC	-02′
	Dip	-03′
	Refraction	-02′
Ho		32°07′

Since it is now past midnight GMT, a new navigational day has started. It is October 2, 1973 and the time is 0200 GMT. (1900 + 7 hours).

We open the Air Almanac (Table VII) and note that for 0200 GMT October 2, 1973 the GHA of Aries is 40°39′. We draw the time diagram (FIG. 32). LHA of Aries is 283°25′.

Since the altitude of Polaris (corrected for the "wobble") is the latitude of the observer, we must look in the Air Almanac for the "Q," (or wobble) correction factor. This information is listed on page A84 (Table VIII). "Q" for LHA 283°25′ = +16′ We add this amount to Ho.

Ho	32°07′
"Q"	+16′

Latitude 32°23′ North

So our sight was 1′ off and our calculated position 1 mile in error. That's good enough for small boat navigation!

Appropriate portions of the "Work Sheet" are used to tabulate the data for this sight (FIG. 33).

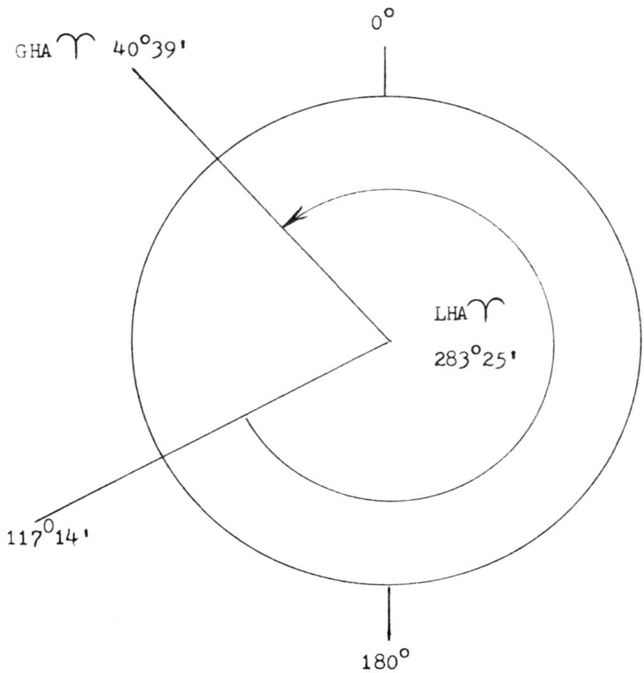

FIGURE 32

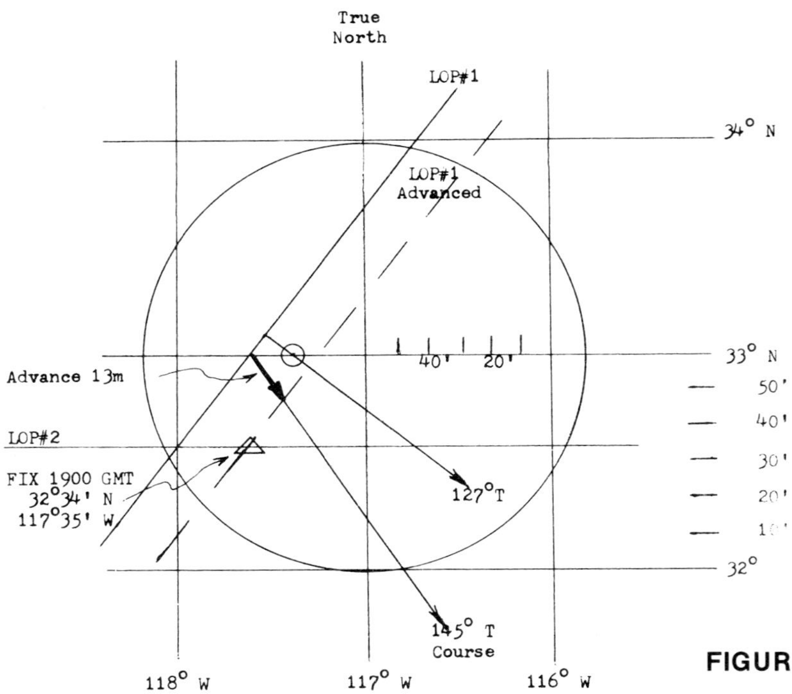

FIGURE 34

ADVANCING A LINE OF POSITION

18

In the practical examples we have worked so far, we have obtained three lines of position. In each case the result showed we were somewhere on a LOP, but the exact spot was not determined. From a single observation of a celestial body, only one LOP can be obtained. To have a "fix," we need at least two LOPs that intersect. The point of intersection of the two LOPs will then determine our position exactly.

Let's go back to our problem in Chapter 15. We took a "sun sight" in the morning and obtained a LOP. We will call this LOP #1 and re-plot it as Figure 34.

To get a second LOP to intersect with LOP#1 we will have to wait until the sun changes azimuth. Furthermore, we ll have to wait a sufficient time for the azimuth to change substantially. This is necessary because if you are going to plot two lines on a chart and determine graphically their point of intersection, the larger the angle between the lines, the less the error is going to be in locating the intersecting point.

Obviously, we are not going to drift around one spot until the sun moves further west so we can get a good second sight. This is what we will do.

We are sailing for the Coronados on a heading of 145°T. Our knotmeter reads a steady 5 knots. Holding this course, we plan to take a second sight on the sun at some later time. LOP#1 was obtained at 17:03:20 GMT. For our second LOP, let's plan a "noon sight" which will be at approximately 1940 GMT. We work out the sight as in Chapter 16, and LOP #2 is determined to be Latitude 32°34′ North.

We draw LOP#2 on the chart (FIG. 34).

LOP#1 was obtained at 1703 GMT or 2 hours 37 minutes earlier. (1940 GMT minus 1703 GMT = 0237). During this time, at our speed of 5 knots, we have traveled 13 miles on a course of 145°T. (2h37m x 5 knots = 13 miles). We now graphically "advance" LOP#1 a distance of 13 miles in the direction 145°T (FIG. 34). Where LOP#2 and LOP#1 ADVANCED intersect, "fixes" our position. We are at 117°35′W - 32°34′N as measured on the plotting chart (FIG. 34).

To obtain a "fix," as mentioned earlier, two or more sights are needed. They can be any combination: sun, moon, planet, "noon," Polaris, or star sights. Often the moon is visible during the day, so a sun and moon sight can be made at practically the same time for an immediate "fix." Often in the early morning and late evening, many star, moon, and planet sights are available at the same time.

The point of intersection, of two LOPs determines a "fix" as we have mentioned (FIG. 35).

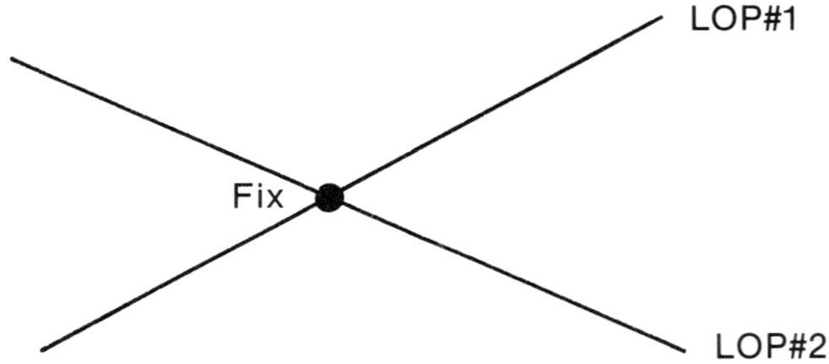

FIGURE 35

If, however, sights are taken on three heavenly bodies, unless absolute accuracy is obtained from the sights, the LOPs, when plotted, will not intersect in a point, but a small triangle will be formed by the three LOPs (FIG. 36).

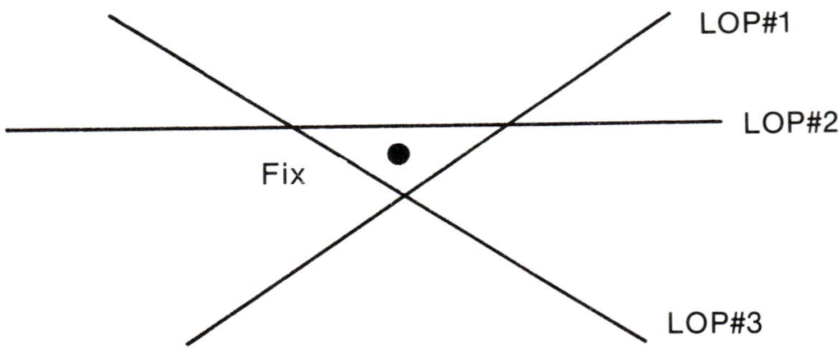

FIGURE 36

In this case the "fix" is considered to be at the center of the triangle. The smaller the triangle, the better the sights.

However, a single LOP is often very valuable to a navigator. If you are sailing up or down a coastline, a single "noon sight" will give you your latitude directly and you will know how far north or south you are. A LOP that is parallel with a coast line will tell you directly how far off the coast you are, and thus be a very valuable "one sight fix."

Knowing a little about the theory of celestial navigation allows you to get the most out of the sights that are available to you.

STAR SIGHTS 19

Star sights are easy to work, using HO249 Volume I. Ho is obtained after correcting Hs, the GHA of Aries is obtained from the Air Almanac in the usual way, and LHA is determined to a whole degree by choosing an appropriate assumed longitude. The time diagram for Aries and your longitude is always drawn.

Opening Volume I to the page for the assumed latitude (Addendum 9), Hc and Zn are read directly on the line for the LHA of Aries, for the particular star listed. Ho and Hc are compared and the LOP is plotted in the usual manner. Note that Zn comes from the table directly.

In Volume I, seven stars are listed for each LHA of Aries, three of which are considered the most suitable for a three-star fix. These three stars are marked by a diamond symbol (♦) in the table.

Volume I is usable only for selected stars, the ones that are listed in the tables.

In making star sights in the morning and in the evening, when the stars and the horizon are both visible, it is advisable to first make an approximate determination of the LHA of Aries at the time the sights are planned. Then, entering Volume I with the latitude and LHA of Aries, the approximate altitudes and azimuths of the listed stars can be determined in advance. This will serve not only in determining what stars are to be used, but it will also assist the observer in finding them in the sextant. Just pre-set the altitude on the sextant to the listed altitude, point the sextant in the direction (Zn) listed, and the star should be in the field of view.

Using this technique, a knowledge of star identification is not actually necessary. But being able to recognize the principal navigation stars is desirable.

A study of the "Star Diagrams" and "Planet Location Diagram" in the Air Almanac is recommended. Various "star-finders" can be purchased, and a study of such books as "Stars," a Golden Nature Guide published by Golden Press of New York, will prove very helpful in shooting the stars, and make it more fun.

After a fix is obtained from "star sights," the fix is adjusted for "Precession and Nutation" by reference to Table 5 in Volume I (Addendum 10). This is a small correction and is the distance and direction the "fix" is moved after it is plotted on the chart.

Precession is the technical term that describes the slow conical motion of the earth's axis relative to the plane of the earth's orbit. Nutation describes the non-uniformity of the precession! Don't worry about it. Just make the small correction listed in Table 5.

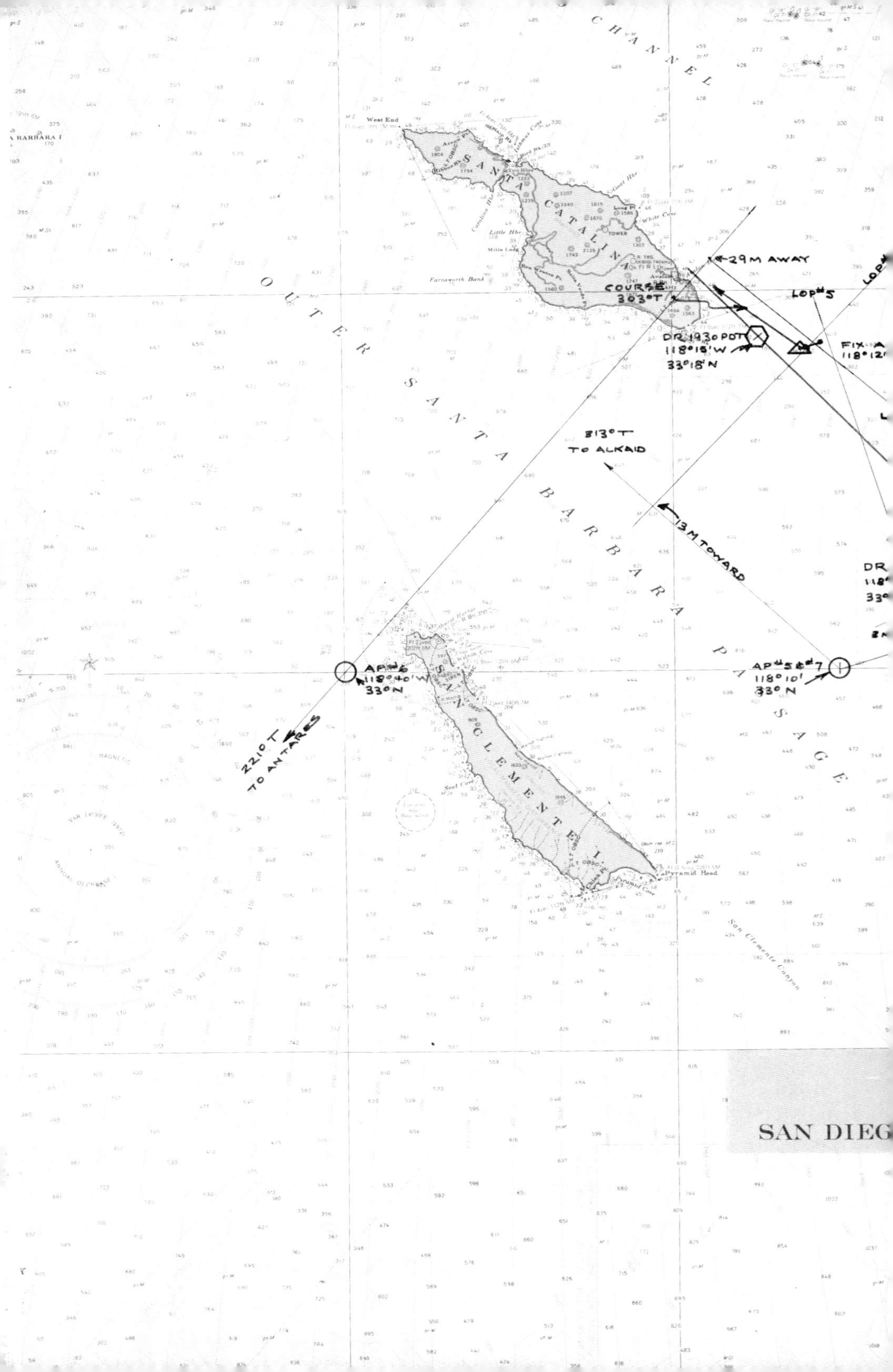

CHANNEL
SANTA CATALINA
West End
Farnsworth Bank
Little Hbr
White Cove
Long Pt
TOWER
COURSE 303°T
29 M AWAY
LOP#5
DR 1930 PDT
118°15'W
33°18'N
OUTER SANTA BARBARA PASSAGE
313°T TO ALKAID
13 M TOWARD
AP#6
118°40'W
33°N
AP#5 & #7
118°10'
33°N
221°T TO ANTARES
SAN CLEMENTE I
Wilson Cove
Seal Cove
China Pt
Pyramid Head
Pyramid Cove
San Clemente Canyon
MAGNETIC
SAN DIEG

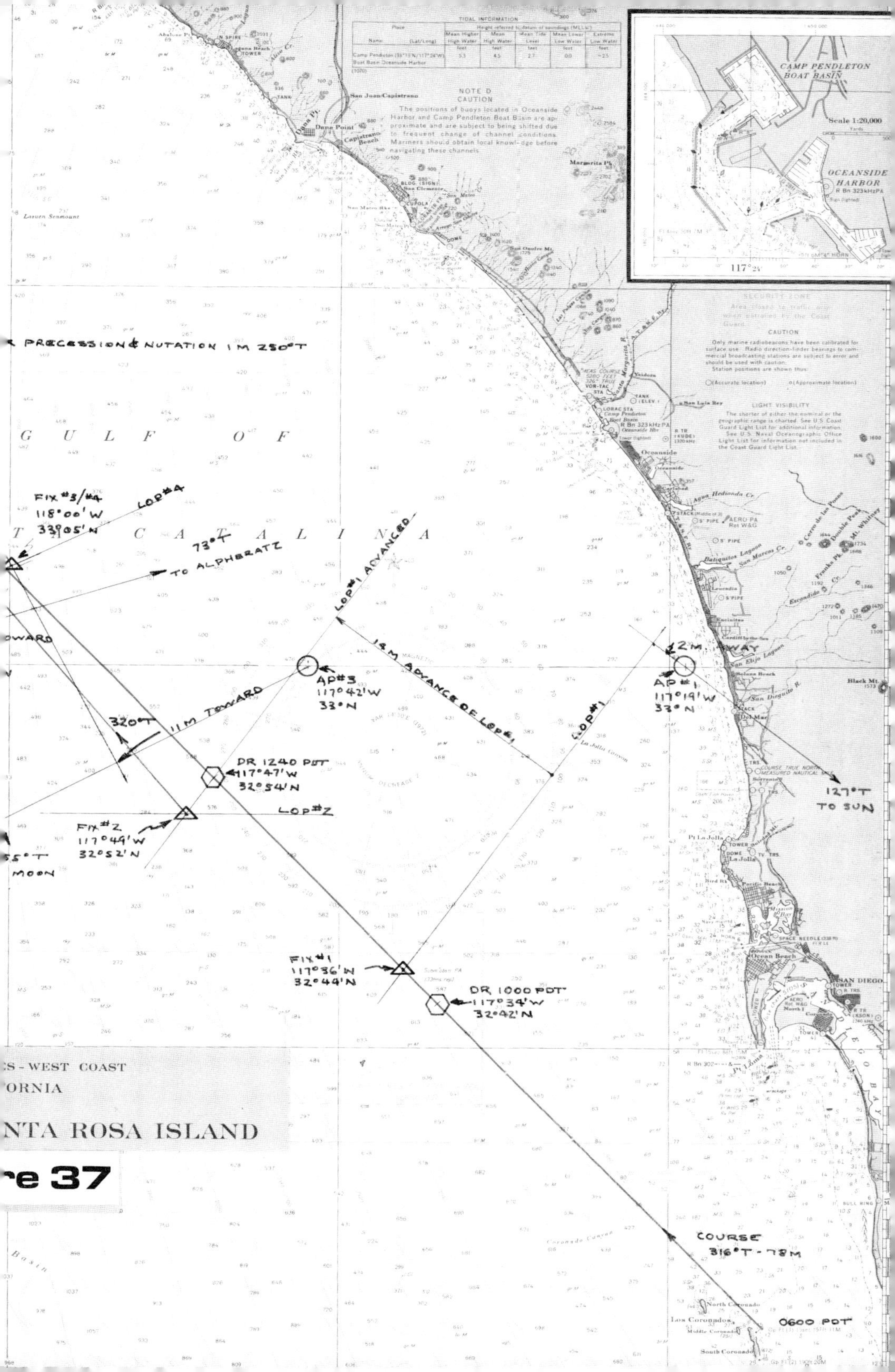
TIDAL INFORMATION
NOTE D
CAUTION
The positions of buoys located in Oceanside Harbor and Camp Pendleton Boat Basin are approximate and are subject to being shifted due to frequent change of channel conditions. Mariners should obtain local knowledge before navigating these channels.
CAMP PENDLETON BOAT BASIN
OCEANSIDE HARBOR
Scale 1:20,000
SECURITY ZONE
CAUTION
Only marine radiobeacons have been calibrated for surface use. Radio direction-finder bearings to commercial broadcasting stations are subject to error and should be used with caution.
LIGHT VISIBILITY
GULF OF
CATALINA
PRECESSION & NUTATION 1M 250°T
FIX #3/#4
118°00'W
33°05'N
LOP#4
73°T
TO ALPHERATZ
LOP#1 ADVANCED
14M ADVANCE OF LOP#1
AP#3
117°42'W
33°N
AP#1
117°19'W
33°N
2M AWAY
LOP#1
127°T
TO SUN
320°T
11M TOWARD
DR 1240 PDT
117°47'W
32°54'N
LOP#2
FIX#2
117°49'W
32°52'N
MOON
FIX#1
117°36'W
32°44'N
DR 1000 PDT
117°34'W
32°42'N
ES - WEST COAST
FORNIA
NTA ROSA ISLAND
re 37
COURSE
316°T - 78M
0600 PDT
Oceanside
San Clemente
Dana Point
La Jolla
Pt La Jolla
SAN DIEGO
Los Coronados
North Coronado
South Coronado

LOG

SOUTH CORONADO TO AVALON

OCTOBER 2, 1973

PDT	
0600	Cleared South Coronado. Course 316°T - 6 knots.
1000	DR: 117°34′W - 32°42′N. Sun sight at 170230 GMT. Fix #1 estimated at 117°36′W - 32°44′N. Knotmeter reading 5 knots.
1240	"Noon Sight" DR: 117°47′W - 32°54′N. Fix #2: 117°49′W - 32°52′N with LOP#1 advanced. Changed course to 320°T. Speed 5 knots.
1600	Sun #3 and Moon #4 sights. DR: 118°01′W - 33°04′N. Fix: 118°00′W - 33°05′N. Changed course to 315°T.
1930	DR Position: 118°15′W - 33°18′N. 3 star fix: Alpheratz #5 - Antares #6 - Alkaid #7; Fix: 118°12′W - 33°17′N. East End light visible off port bow, Avalon lights ahead. Changed course to 303°T - Speed 6 knots.
2100	Picked up mooring in Descanso Bay.

FIGURE 38

WORK SHEETS AND A TYPICAL CRUISE 20

The practical aspects of a navigator's job can best be demonstrated by taking a typical cruise and working out the navigational problems involved.

In Chapter 17 we said that we had "dropped the hook" in the lee of South Coronado Island in the late afternoon of October 1, 1973. We now plan to leave in the early morning of October 2, 1973 and sail a direct course to Avalon Bay, Santa Catalina Island. Pre-planning the trip, we plot the course on C&GS Chart 5101 (FIG. 37) and determine the heading will be 316°T. The distance is 78 nautical miles. At 5 knots, the trip will take a little over fifteen and a half hours.

To illustrate how celestial navigation can be used on such a trip, we will plan a series of "sights" far more comprehensive than needed or would be used on an actual trip of this type. They will, however, demonstrate techniques available to the navigator.

We will plan to leave the Coronados around 0600 PDT, take a sun sight (#1) around 1000 PDT, a "noon sight" (#2) around 1240 PDT, sun (#3) and moon (#4) sights around 1600 PDT as the moon will be visible at that time, and a series of three star sights at twilight around 1900 PDT (#5, #6, #7). If all goes as planned, we'll arrive at Avalon and anchor off Casino Point around 2130 PDT.

The information contained in the log of the cruise (FIG. 38), and the Work Sheets (FIG. 39 thru 45) will be easily followed. In this exercise we'll do all of the plotting directly on C&GS Chart 5101 (FIG.37). The pages needed from the Air Almanac and HO249 Volumes I & II, are reproduced in the Addenda for reference.

WORK SHEET #1

Body SUN Date 2 OCT 73

Time of sight 17:02:30 GMT

GHA(*) SUN at 1700 GMT 77°41'

Increment for 2:30 38'

GHA(*) SUN 78°19'

DR (Long. 117°34' W
(Lat. 32°42' N

(*) GHA of Body, or
GHA of Aries for star sight

Declination S 3°43'

Hs 37°45'

SC ~~0~~

IC −2'

Dip −3'

Refr −1'

SD +16'

P in A —

Ho 37°55'

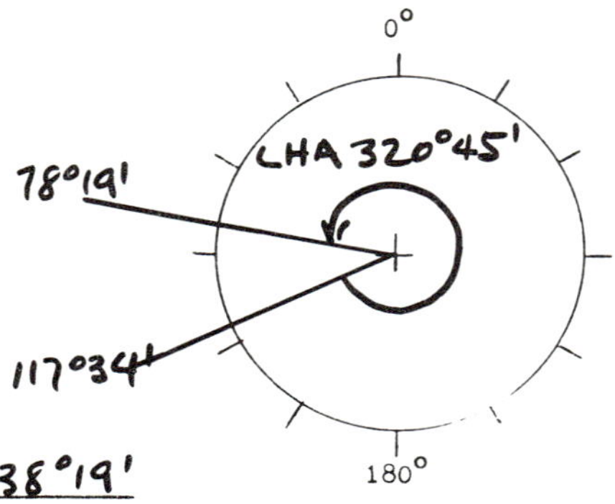

TIME DIAGRAM

GHA(*) SUN 78°19' + (360°) 438°19'

Assumed Longitude 117°19'

LHA 321°

Assumed Latitude 33°N

Declination Argument 3° ~~(Same)~~ (Contrary)

HO249 Hc 38°29' d −44 Z 127°

d correction −32'

Hc 37°57' Zn 127°T

Hc - Ho +2 (Away) ~~(Toward)~~
+ −

NOON SIGHT

Latitude = ZD ± Declination

89° 60'

(-Ho) ________

(=ZD) ________

(±Dec) ________

Latitude = ________

POLARIS SIGHT

Ho ________

Q for LHA ♈ ________ ________

Latitude ________

FIGURE 39

WORK SHEET #2

Body SUN Date 2 OCT 73

Time of sight 1900 + GMT

GHA(*) SUN at NOON GMT ______

Increment for ______ ______

GHA(*) ______ ______

Hs 53°13'

SC -0-

IC -2'

Dip -3'

Refr -1'

SD +16'

P in A —

Ho 53°23'

GHA(*) ______ ______ + (360°) ______

Assumed Longitude ______

LHA ______

Assumed Latitude ______

Declination Argument ______ (Same) (Contrary)

HO249 Hc ______ d ______ Z ______

d correction ______

Hc ______ Zn ______

Hc - Ho ______ (Away) + (Toward) -

DR (Long. 117°47' W
(Lat. 32°54' N

(*) GHA of Body, or
GHA of Aries for star sight

Declination S 3°45'

0°

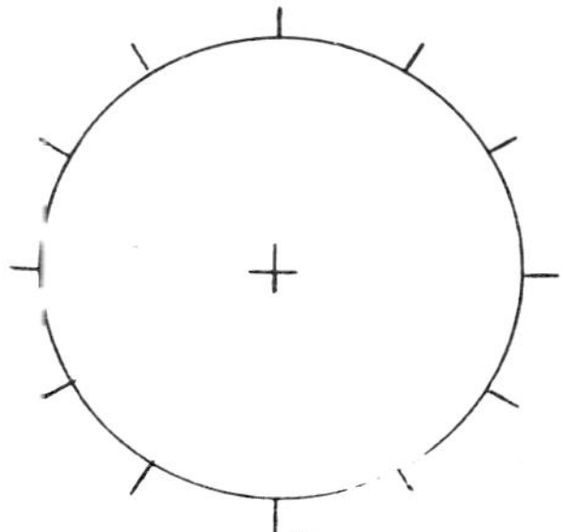

180°

TIME DIAGRAM

NOON SIGHT

Latitude = ZD ± Declination

89° 60'

(-Ho) 53°23'

(=ZD) 36°37'

(±Dec) - 3°45'

Latitude = 32°52' N

POLARIS SIGHT

Ho ______

Q for LHA ♈ ______ ______

Latitude ______

FIGURE 40

WORK SHEET #3

Body SUN Date 2 OCT 73

Time of sight 23:00:00 GMT

GHA(*) SUN at 2300 GMT 167°42'

Increment for —

GHA(*) SUN 167°42'

DR (Long. 118°01' W

DR (Lat. 33°04'N

(*) GHA of Body, or
GHA of Aries for star sight

Declination S 3°48'

Hs 30°09'

SC ~~0~~

IC -2'

Dip -3'

Refr -2'

SD +16'

P in A —

Ho 30°18'

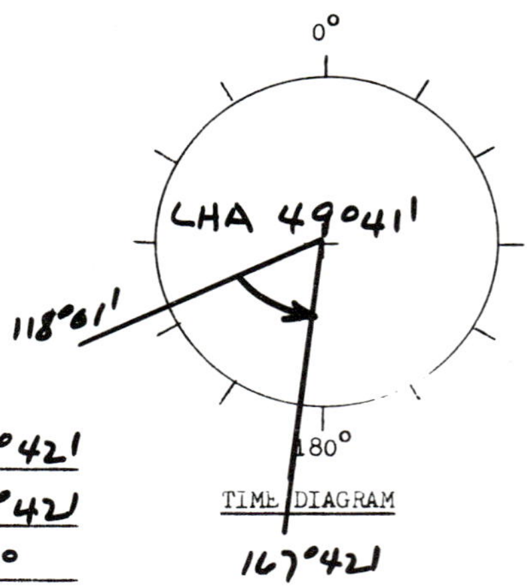

TIME DIAGRAM

GHA(*) SUN 167°42' + ~~(360°)~~ 167°42'

Assumed Longitude 117°42'

LHA 50°

Assumed Latitude 33° N

Declination Argument 3° ~~(Same)~~ (Contrary)

HO249 Hc 30°39' d -40 Z 117°

d correction -32'

Hc 30°07'

Zn 243° T

360°
-117°
243°

Hc - Ho -11 ~~(Away)~~ + (Toward) -

NOON SIGHT

Latitude = ZD ± Declination

89° 60'

(-Ho) ______

(=ZD) ______

(±Dec) ______

Latitude = ______

POLARIS SIGHT

Ho ______

Q for LHA ♈ ______ ______

Latitude ______

FIGURE 41

WORK SHEET #4

Body MOON Date 2 OCT 73

Time of sight 23:02:10 GMT

GHA(*) ______ at 2300 GMT 93°33'

Increment for 2:10 +32

GHA(*) MOON 94°05'

DR (Long. 118°01' W
(Lat. 33°04' N

(*) GHA of Body, or
GHA of Aries for star sight

Declination S 24°03'

Hs 27°33'

SC —0—

IC -2'

Dip -3'

Refr -2'

SD +15'

P in A +48'

Ho 28°29'

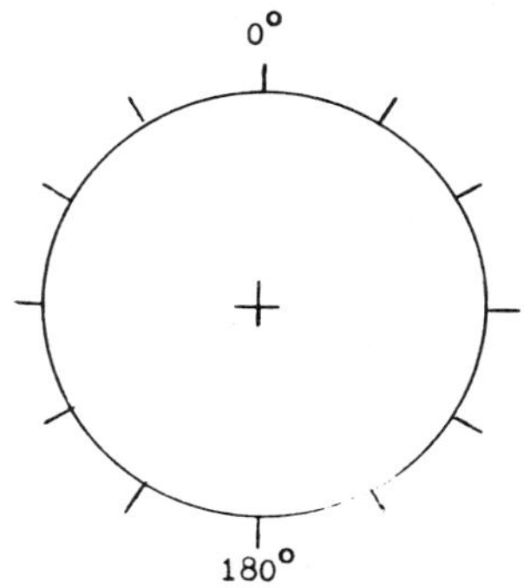

TIME DIAGRAM

GHA(*) MOON 94°05' + (360°) 454°05'

Assumed Longitude 118°05'

LHA 336°

Assumed Latitude 33°N

Declination Argument 24° ~~(Same)~~ (Contrary)

HO249 Hc 28°35' d -55 Z 155°

d correction -03

Hc 28°32' Zn 155°T

Hc - Ho +3 (Away) ~~(Toward)~~
+ -

NOON SIGHT

Latitude = ZD ± Declination

89° 60'

(-Ho) ______

(=ZD) ______

(±Dec) ______

Latitude = ______

POLARIS SIGHT

Ho ______

Q for LHA ♈ ______ ______

Latitude ______

FIGURE 42

WORK SHEET #5

Body ALPHERATZ Date 3 OCT 73   DR (Long. 118°15'W (Lat. 33°18'N

Time of sight 02:30:00 GMT

GHA(*) ♈ at 0230 GMT 49°10'

(*) GHA of Body, or
GHA of Aries for star sight

Increment for — ______

GHA(*) ♈ 49°10'

Declination ______

Hs 30°33'

SC —0—

IC −2'

Dip −3'

Refr −2'

SD —

P in A —

Ho 30°26'

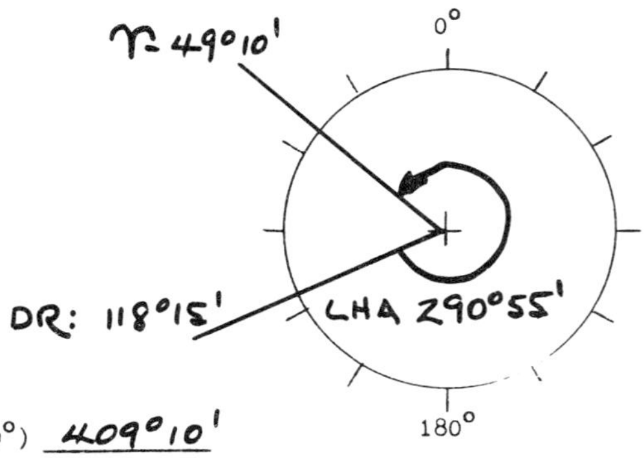

TIME DIAGRAM

GHA(*) ARIES 49°10' + (360°) 409°10'

Assumed Longitude 118°10'

LHA 291°

Assumed Latitude 33°N

Declination Argument ______ (Same) (Contrary)

HO249 Hc ______ d ______ Z ______

d correction ______

Hc 30°21'   Zn 073°T

Hc - Ho −5 ~~(Away)~~ + (Toward) −

NOON SIGHT

Latitude = ZD ± Declination

89° 60'

(−Ho) ______

(=ZD) ______

(±Dec) ______

Latitude = ______

POLARIS SIGHT

Ho ______

Q for LHA ♈ ______ ______

Latitude ______

PRECESSION & NUTATION CORRECTION 1M − 250°

FIGURE 43

WORK SHEET #6

Body ANTARES Date 3 OCT 73

Time of sight 02:32:00 GMT

GHA(*) ♈ at 0230 GMT 49°10'

Increment for 2:00 30'

GHA(*) ♈ 49°40'

Hs 17°00'

SC -0-

IC -2'

Dip -3'

Refr -3'

SD —

P in A —

Ho 16°52'

GHA(*) ARIES 49°40' + (360°) 409°40'

Assumed Longitude 118°40'

LHA 291°

Assumed Latitude 33°N

Declination Argument ______ (Same) (Contrary)

HO249 Hc ______ d ______ Z ______

d correction ______

Hc 17°21' Zn 221°T

Hc - Ho +29 (Away) + ~~(Toward)~~ -

DR (Long. 118°15'W

DR (Lat. 33°18'N

(*) GHA of Body, or
GHA of Aries for star sight

Declination ______

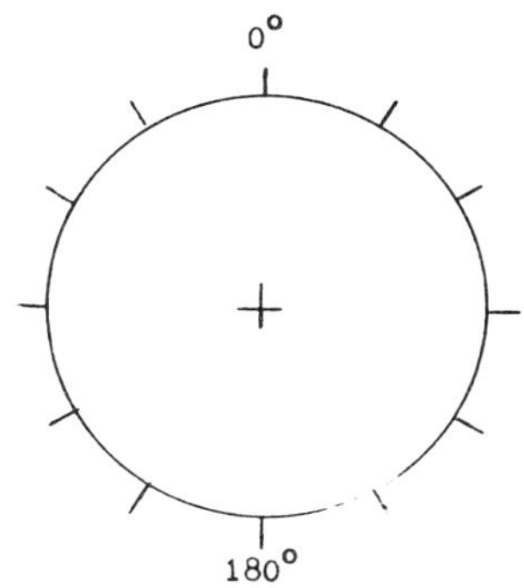

TIME DIAGRAM

NOON SIGHT

Latitude = ZD ± Declination

89° 60'

(-Ho) ______

(=ZD) ______

(±Dec) ______

Latitude = ______

POLARIS SIGHT

Ho ______

Q for LHA ♈ ______ ______

Latitude ______

FIGURE 44

WORK SHEET #7

Body ALKAID Date 3 OCT 73

DR (Long. 118°15' W (Lat. 33°18' N

Time of sight 02:34:00 GMT

GHA(*) ♈ at 0230 GMT 49°10'

(*) GHA of Body, or GHA of Aries for star sight

Increment for 4:00 1°00'

GHA(*) ♈ 50°10'

Declination ____

Hs 27°35'

SC ~~0~~

IC −2'

Dip −3'

Refr −2'

SD —

P in A —

Ho 27°28'

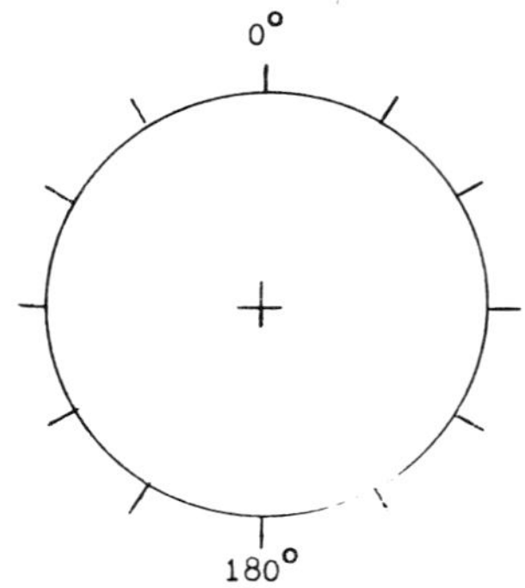

TIME DIAGRAM

GHA(*) ARIES 50°10' + (360°) 410°10'

Assumed Longitude 118°10'

LHA 292°

Assumed Latitude 33°N

Declination Argument ____ (Same) (Contrary)

HO249 Hc ____ d ____ Z ____

d correction ____

Hc 27°15'

Zn 313°T

Hc - Ho −13 ~~(Away) +~~ (Toward) −

NOON SIGHT

Latitude = ZD ± Declination

89° 60'

(-Ho) ____

(=ZD) ____

(±Dec) ____

Latitude = ____

POLARIS SIGHT

Ho ____

Q for LHA ♈ ____ ____

Latitude ____

FIGURE 45

CORRECTIONS TO BE APPLIED TO MARINE SEXTANT ALTITUDES

MARINE SEXTANT ERROR

Sextant Number

Index Error

CORRECTIONS

In addition to sextant error and dip, corrections are to be applied for:
- Refraction
- Semi-diameter (for the Sun and Moon)
- Parallax (for the Moon)
- Dome refraction (if applicable)

CORRECTION FOR DIP OF THE HORIZON

To be subtracted from sextant altitude

Ht.	Dip	Ht.	Dip	Ht.	Dip	Ht.	Dip	Ht.	Dip
Ft.	′	Ft.	′	Ft.	′	Ft.	′	Ft.	′
0		114		437		968		1 707	
	1		11		21		31		41
2		137		481		1 033		1 792	
	2		12		22		32		42
6		162		527		1 099		1 880	
	3		13		23		33		43
12		189		575		1 168		1 970	
	4		14		24		34		44
21		218		625		1 239		2 061	
	5		15		25		35		45
31		250		677		1 311		2 155	
	6		16		26		36		46
43		283		731		1 386		2 251	
	7		17		27		37		47
58		318		787		1 463		2 349	
	8		18		28		38		48
75		356		845		1 543		2 449	
	9		19		29		39		49
93		395		906		1 624		2 551	
	10		20		30		40		50
114		437		968		1 707		2 655	

TABLE I (Addendum 4) Dip Correction Air Almanac Back Cover

CORRECTIONS TO BE APPLIED TO SEXTANT ALTITUDE

REFRACTION

To be subtracted from sextant altitude (referred to as observed altitude in A.P. 3270)

$R_\circ$	Height above sea level in units of 1 000 ft.												$R_\circ$	$R = R_\circ \times f$			
	0	5	10	15	20	25	30	35	40	45	50	55		*f*			
	Sextant Altitude													0·9	1·0	1·1	1·2
														R			
′	° ′	° ′	° ′	° ′	° ′	° ′	° ′	° ′	° ′	° ′	° ′	° ′	′	′	′	′	′
	90	90	90	90	90	90	90	90	90	90	90	90					
0													0	0	0	0	0
	63	59	55	51	46	41	36	31	25	20	17	13					
1													1	1	1	1	1
	33	29	26	22	19	16	14	11	9	7	6	4					
2													2	2	2	2	2
	21	19	16	14	12	10	8	7	5	4	2 40	1 40					
3													3	3	3	3	4
	16	14	12	10	8	7	6	5	3 10	2 20	1 30	0 40					
4													4	4	4	4	5
	12	11	9	8	7	5	4 00	3 10	2 10	1 30	0 39	+0 05					
5													5	5	5	5	6
	10	9	7	5 50	4 50	3 50	3 10	2 20	1 30	0 49	+0 11	−0 19					
6													6	5	6	7	7
	8 10	6 50	5 50	4 50	4 00	3 00	2 20	1 50	1 10	0 24	−0 11	−0 38					
7													7	6	7	8	8
	6 50	5 50	5 00	4 00	3 10	2 30	1 50	1 20	0 38	+0 04	−0 28	−0 54					
8													8	7	8	9	10
	6 00	5 10	4 10	3 20	2 40	2 00	1 30	1 00	0 19	−0 13	−0 42	−1 08					
9													9	8	9	10	11
	5 20	4 30	3 40	2 50	2 10	1 40	1 10	0 35	+0 03	−0 27	−0 53	−1 18					
10													10	9	10	11	12
	4 30	3 40	2 50	2 20	1 40	1 10	0 37	+0 11	−0 16	−0 43	−1 08	−1 31					
12													12	11	12	13	14
	3 30	2 50	2 10	1 40	1 10	0 34	+0 09	−0 14	−0 37	−1 00	−1 23	−1 44					
14													14	13	14	15	17
	2 50	2 10	1 40	1 10	0 37	+0 10	−0 13	−0 34	−0 53	−1 14	−1 35	−1 56					
16													16	14	16	18	19
	2 20	1 40	1 20	0 43	+0 15	−0 08	−0 31	−0 52	−1 08	−1 27	−1 46	−2 05					
18													18	16	18	20	22
	1 50	1 20	0 49	+0 23	−0 02	−0 26	−0 46	−1 06	−1 22	−1 39	−1 57	−2 14					
20													20	18	20	22	24
	1 12	0 44	+0 19	−0 06	−0 28	−0 48	−1 09	−1 27	−1 42	−1 58	−2 14	−2 30					
25													25	22	25	28	30
	0 34	+0 10	−0 13	−0 36	−0 55	−1 14	−1 32	−1 51	−2 06	−2 21	−2 34	−2 49					
30													30	27	30	33	36
	+0 06	−0 16	−0 37	−0 59	−1 17	−1 33	−1 51	−2 07	−2 23	−2 37	−2 51	−3 04					
35													35	31	35	38	42
	−0 18	−0 37	−0 58	−1 16	−1 34	−1 49	−2 06	−2 22	−2 35	−2 49	−3 03	−3 16					
40													40	36	40	44	48
		−0 53	−1 14	−1 31	−1 47	−2 03	−2 18	−2 33	−2 47	−2 59	−3 13	−3 25					
45													45	40	45	50	54
		−1 10	−1 28	−1 44	−1 59	−2 15	−2 28	−2 43	−2 56	−3 08	−3 22	−3 33					
50													50	45	50	55	60
			−1 40	−1 53	−2 09	−2 24	−2 38	−2 52	−3 04	−3 17	−3 29	−3 41					
55													55	49	55	60	66
				−2 03	−2 18	−2 33	−2 46	−3 01	−3 12	−3 25	−3 37	−3 48					
60													60	54	60	66	72
							−2 53	−3 07	−3 19	−3 31	−3 42	−3 53					

TABLE II (Addendum 5) Refraction Correction
Air Almanac Inside Back Cover

548 (DAY 274) GREENWICH P. M. 1973 OCTOBER 1 (MONDAY)

GMT	☉ SUN GHA	SUN Dec.	ARIES GHA ♈	VENUS–3.7 GHA	VENUS Dec.	MARS–1.9 GHA	MARS Dec.	JUPITER–2.1 GHA	JUPITER Dec.	◐ MOON GHA	MOON Dec.
h m	° ′	° ′	° ′	° ′	° ′	° ′	° ′	° ′	° ′	° ′	° ′
12 00	2 34.7	S 3 14.4	190 04.7	321 57	S19 42	153 10	N11 23	245 19	S20 24	306 12	S23 33
10	5 04.8	14.5	192 35.1	324 27		155 41		247 50		308 37	33
20	7 34.8	14.7	195 05.5	326 57		158 11		250 20		311 02	34
30	10 04.8	· 14.8	197 35.9	329 27	·	160 42	·	252 50	·	313 27	· 34
40	12 34.9	15.0	200 06.3	331 57		163 12		255 21		315 52	34
50	15 04.9	15.2	202 36.7	334 27		165 43		257 51		318 17	35
13 00	17 34.9	S 3 15.3	205 07.1	336 57	S19 43	168 13	N11 23	260 22	S20 24	320 42	S23 35
10	20 05.0	15.5	207 37.5	339 26		170 44		262 52		323 07	36
20	22 35.0	15.6	210 07.9	341 56		173 14		265 22		325 32	36
30	25 05.0	· 15.8	212 38.4	344 26	·	175 45	·	267 53	·	327 57	· 37
40	27 35.1	16.0	215 08.8	346 56		178 15		270 23		330 22	37
50	30 05.1	16.1	217 39.2	349 26		180 46		272 54		332 47	38
14 00	32 35.1	S 3 16.3	220 09.6	351 56	S19 44	183 16	N11 23	275 24	S20 24	335 12	S23 38
10	35 05.2	16.5	222 40.0	354 26		185 46		277 54		337 37	39
20	37 35.2	16.6	225 10.4	356 56		188 17		280 25		340 02	39
30	40 05.2	· 16.8	227 40.8	359 26	·	190 47	·	282 55	·	342 27	· 40
40	42 35.3	16.9	230 11.2	1 56		193 18		285 26		344 51	40
50	45 05.3	17.1	232 41.6	4 26		195 48		287 56		347 16	40
15 00	47 35.3	S 3 17.3	235 12.1	6 56	S19 45	198 19	N11 23	290 26	S20 24	349 41	S23 41
10	50 05.4	17.4	237 42.5	9 26		200 49		292 57		352 06	41
20	52 35.4	17.6	240 12.9	11 55		203 20		295 27		354 31	42
30	55 05.4	· 17.7	242 43.3	14 25	·	205 50	·	297 58	·	356 56	· 42
40	57 35.5	17.9	245 13.7	16 55		208 21		300 28		359 21	43
50	60 05.5	18.1	247 44.1	19 25		210 51		302 58		1 46	43
16 00	62 35.5	S 3 18.2	250 14.5	21 55	S19 46	213 22	N11 23	305 29	S20 24	4 11	S23 43
10	65 05.6	18.4	252 44.9	24 25		215 52		307 59		6 36	44
20	67 35.6	18.6	255 15.3	26 55		218 23		310 30		9 01	44
30	70 05.6	· 18.7	257 45.8	29 25	·	220 53	·	313 00	·	11 26	· 45
40	72 35.7	18.9	260 16.2	31 55		223 24		315 30		13 51	45
50	75 05.7	19.0	262 46.6	34 25		225 54		318 01		16 16	45
17 00	77 35.7	S 3 19.2	265 17.0	36 55	S19 47	228 25	N11 23	320 31	S20 24	18 41	S23 46
10	80 05.8	19.4	267 47.4	39 25		230 55		323 02		21 06	46
20	82 35.8	19.5	270 17.8	41 55		233 26		325 32		23 30	47
30	85 05.8	· 19.7	272 48.2	44 25	·	235 56	·	328 03	·	25 55	· 47
40	87 35.9	19.8	275 18.6	46 54		238 27		330 33		28 20	47
50	90 05.9	20.0	277 49.0	49 24		240 57		333 03		30 45	48
18 00	92 35.9	S 3 20.2	280 19.4	51 54	S19 48	243 28	N11 23	335 34	S20 24	33 10	S23 48
10	95 06.0	20.3	282 49.9	54 24		245 58		338 04		35 35	48
20	97 36.0	20.5	285 20.3	56 54		248 29		340 35		38 00	49
30	100 06.0	· 20.7	287 50.7	59 24	·	250 59	·	343 05	·	40 25	· 49
40	102 36.1	20.8	290 21.1	61 54		253 30		345 35		42 50	49
50	105 06.1	21.0	292 51.5	64 24		256 00		348 06		45 15	50
19 00	107 36.1	S 3 21.1	295 21.9	66 54	S19 49	258 31	N11 23	350 36	S20 24	47 40	S23 50
10	110 06.2	21.3	297 52.3	69 24		261 01		353 07		50 05	51
20	112 36.2	21.5	300 22.7	71 54		263 32		355 37		52 30	51
30	115 06.2	· 21.6	302 53.1	74 24	·	266 02	·	358 07	·	54 55	· 51
40	117 36.3	21.8	305 23.6	76 54		268 33		0 38		57 20	52
50	120 06.3	21.9	307 54.0	79 23		271 03		3 08		59 45	52
20 00	122 36.3	S 3 22.1	310 24.4	81 53	S19 50	273 34	N11 23	5 39	S20 24	62 10	S23 52
10	125 06.4	22.3	312 54.8	84 23		276 04		8 09		64 35	53
20	127 36.4	22.4	315 25.2	86 53		278 34		10 39		66 59	53
30	130 06.4	· 22.6	317 55.6	89 23	·	281 05	·	13 10	·	69 24	· 53
40	132 36.5	22.8	320 26.0	91 53		283 35		15 40		71 49	54
50	135 06.5	22.9	322 56.4	94 23		286 06		18 11		74 14	54
21 00	137 36.5	S 3 23.1	325 26.8	96 53	S19 50	288 36	N11 23	20 41	S20 24	76 39	S23 54
10	140 06.6	23.2	327 57.2	99 23		291 07		23 11		79 04	54
20	142 36.6	23.4	330 27.7	101 53		293 37		25 42		81 29	55
30	145 06.6	· 23.6	332 58.1	104 23	·	296 08	·	28 12	·	83 54	· 55
40	147 36.7	23.7	335 28.5	106 53		298 38		30 43		86 19	55
50	150 06.7	23.9	337 58.9	109 23		301 09		33 13		88 44	56
22 00	152 36.7	S 3 24.0	340 29.3	111 53	S19 51	303 39	N11 23	35 44	S20 24	91 09	S23 56
10	155 06.8	24.2	342 59.7	114 23		306 10		38 14		93 34	56
20	157 36.8	24.4	345 30.1	116 52		308 40		40 44		95 59	57
30	160 06.8	· 24.5	348 00.5	119 22	·	311 11	·	43 15	·	98 24	· 57
40	162 36.9	24.7	350 30.9	121 52		313 41		45 45		100 49	57
50	165 06.9	24.9	353 01.4	124 22		316 12		48 16		103 14	57
23 00	167 36.9	S 3 25.0	355 31.8	126 52	S19 52	318 42	N11 23	50 46	S20 24	105 38	S23 58
10	170 07.0	25.2	358 02.2	129 22		321 13		53 16		108 03	58
20	172 37.0	25.3	0 32.6	131 52		323 43		55 47		110 28	58
30	175 07.0	· 25.5	3 03.0	134 22	·	326 14	·	58 17	·	112 53	· 5[illegible]
40	177 37.1	25.7	5 33.4	136 52		328 44		60 48		115 18	59
50	180 07.1	25.8	8 03.8	139 22		331 15		63 18		117 43	59
Rate	15 00.2	S0 01.0		14 59.6	S0 00.9	15 02.9	S0 00.1	15 02.4	0 00.0	14 29.6	S0 02.2*

Lat.	Moon-set	Diff.
N		
°	h m	m
72	■	*
70	■	*
68	■	*
66	16 13	*
64	17 26	15
62	18 02	18
60	18 28	20
58	18 48	21
56	19 05	21
54	19 19	22
52	19 32	22
50	19 43	22
45	20 06	23
40	20 24	24
35	20 40	24
30	20 53	24
20	21 16	24
10	21 36	25
0	21 55	25
10	22 13	26
20	22 33	26
30	22 56	26
35	23 10	26
40	23 25	27
45	23 44	27
50	24 07	27
52	24 18	27
54	24 31	28
56	24 45	28
58	25 02	28
60	00 18	36
S		

Moon's P. in A.

Alt. °	Corr. + ′	Alt. °	Corr. + ′
0	55	56	30
4	54	57	29
11	53	58	28
16	52	59	27
19	51	60	26
22	50	62	25
25	49	63	24
27	48	64	23
29	47	65	22
31	46	66	21
33	45	67	20
35	44	69	19
37	43	70	18
38	42	71	17
40	41	72	16
42	40	73	15
43	39	74	14
45	38	75	13
46	37	76	12
48	36	77	11
49	35	78	10
50	34	79	
52	33		
53	32		
54	31		
56	30		
57			

Sun SD 16.′0

Moon SD 15′

Age 5d

TABLE III Air Almanac page 548 Greenwich P.M. 1973 October 1

INTERPOLATION OF G.H.A.

Increment to be added for intervals of G.M.T. to G.H.A. of: Sun, Aries (♈) and planets; Moon

SUN, etc. m s	° ′	MOON m s	SUN, etc. m s	° ′	MOON m s	SUN, etc. m s	° ′	MOON m s
00 00	0 00	00 00	03 17	0 50	03 25	06 37	1 40	06 52
01	0 01	00 02	21	0 51	03 29	41	1 41	06 56
05	0 02	00 06	25	0 52	03 33	45	1 42	07 00
09	0 03	00 10	29	0 53	03 37	49	1 43	07 04
13	0 04	00 14	33	0 54	03 41	53	1 44	07 08
17	0 05	00 18	37	0 55	03 45	06 57	1 45	07 13
21	0 06	00 22	41	0 56	03 49	07 01	1 46	07 17
25	0 07	00 26	45	0 57	03 54	05	1 47	07 21
29	0 08	00 31	49	0 58	03 58	09	1 48	07 25
33	0 09	00 35	53	0 59	04 02	13	1 49	07 29
37	0 10	00 39	03 57	1 00	04 06	17	1 50	07 33
41	0 11	00 43	04 01	1 01	04 10	21	1 51	07 37
45	0 12	00 47	05	1 02	04 14	25	1 52	07 42
49	0 13	00 51	09	1 03	04 19	29	1 53	07 46
53	0 14	00 55	13	1 04	04 23	33	1 54	07 50
00 57	0 15	01 00	17	1 05	04 27	37	1 55	07 54
01 01	0 16	01 04	21	1 06	04 31	41	1 56	07 58
05	0 17	01 08	25	1 07	04 35	45	1 57	08 02
09	0 18	01 12	29	1 08	04 39	49	1 58	08 06
13	0 19	01 16	33	1 09	04 43	53	1 59	08 11
17	0 20	01 20	37	1 10	04 48	07 57	2 00	08 15
21	0 21	01 24	41	1 11	04 52	08 01	2 01	08 19
25	0 22	01 29	45	1 12	04 56	05	2 02	08 23
29	0 23	01 33	49	1 13	05 00	09	2 03	08 27
33	0 24	01 37	53	1 14	05 04	13	2 04	08 31
37	0 25	01 41	04 57	1 15	05 08	17	2 05	08 35
41	0 26	01 45	05 01	1 16	05 12	21	2 06	08 40
45	0 27	01 49	05	1 17	05 17	25	2 07	08 44
49	0 28	01 53	09	1 18	05 21	29	2 08	08 48
53	0 29	01 58	13	1 19	05 25	33	2 09	08 52
01 57	0 30	02 02	17	1 20	05 29	37	2 10	08 56
02 01	0 31	02 06	21	1 21	05 33	41	2 11	09 00
05	0 32	02 10	25	1 22	05 37	45	2 12	09 04
09	0 33	02 14	29	1 23	05 41	49	2 13	09 09
13	0 34	02 18	33	1 24	05 46	53	2 14	09 13
17	0 35	02 22	37	1 25	05 50	08 57	2 15	09 17
21	0 36	02 27	41	1 26	05 54	09 01	2 16	09 21
25	0 37	02 31	45	1 27	05 58	05	2 17	09 25
29	0 38	02 35	49	1 28	06 02	09	2 18	09 29
33	0 39	02 39	53	1 29	06 06	13	2 19	09 33
37	0 40	02 43	05 57	1 30	06 10	17	2 20	09 38
41	0 41	02 47	06 01	1 31	06 15	21	2 21	09 42
45	0 42	02 51	05	1 32	06 19	25	2 22	09 46
49	0 43	02 56	09	1 33	06 23	29	2 23	09 50
53	0 44	03 00	13	1 34	06 27	33	2 24	09 54
02 57	0 45	03 04	17	1 35	06 31	37	2 25	09 58
03 01	0 46	03 08	21	1 36	06 35	41	2 26	10 00
05	0 47	03 12	25	1 37	06 39	45	2 27	
09	0 48	03 16	29	1 38	06 44	49	2 28	
13	0 49	03 20	33	1 39	06 48	53	2 29	
17	0 50	03 25	37	1 40	06 52	09 57	2 30	
03 21		03 29	06 41		06 56	10 00		

TABLE IV (Addendum 3)

Interpolation of G.H.A.
Air Almanac Inside Front Cover

LAT 33°

DECLINATION (0°-14°) CONTRARY NAME TO LATITUDE

N. Lat. {LHA greater than 180° .... Zn=Z; LHA less than 180° .... Zn=360−Z}

LHA	0° Hc	d	Z	1° Hc	d	Z	2° Hc	d	Z	3° Hc	d	Z	4° Hc	d	Z
	° ′	′	°	° ′	′	°	° ′	′	°	° ′	′	°	° ′	′	°
69	17 29	34	102	16 55	34	103	16 21	35	104	15 46	35	104	15 11	35	105
68	18 19	35	102	17 44	35	103	17 09	35	104	16 34	35	105	15 59	35	106
67	19 08	35	103	18 33	35	104	17 58	35	105	17 23	35	106	16 48	36	106
66	19 57	35	104	19 22	35	105	18 47	36	105	18 11	35	106	17 36	36	107
65	20 46	−35	104	20 11	−36	105	19 35	−35	106	19 00	−36	107	18 24	−36	108
64	21 34	35	105	20 59	36	106	20 23	35	107	19 48	36	108	19 12	37	108
63	22 23	36	106	21 47	35	106	21 12	36	107	20 36	37	108	19 59	36	109
62	23 11	35	106	22 36	36	107	22 00	37	108	21 23	36	109	20 47	37	110
61	24 00	36	107	23 24	37	108	22 47	36	109	22 11	37	109	21 34	37	110
60	24 48	−37	108	24 11	−36	108	23 35	−37	109	22 58	−37	110	22 21	−37	111
59	25 36	37	108	24 59	37	109	24 22	37	110	23 45	37	111	23 08	37	112
58	26 23	36	109	25 47	37	110	25 10	38	111	24 32	37	111	23 55	38	112
57	27 11	37	110	26 34	37	110	25 57	38	111	25 19	38	112	24 41	38	113
56	27 58	37	110	27 21	38	111	26 43	37	112	26 06	38	113	25 28	39	113
55	28 45	−37	111	28 08	−38	112	27 30	−38	113	26 52	−39	114	26 13	−38	114
54	29 32	38	112	28 54	38	113	28 16	38	113	27 38	39	114	26 59	39	115
53	30 19	38	112	29 41	39	113	29 02	38	114	28 24	39	115	27 45	40	116
52	31 05	38	113	30 27	39	114	29 48	39	115	29 09	39	116	28 30	40	117
51	31 51	38	114	31 13	39	115	30 34	40	116	29 54	39	117	29 15	40	117
50	32 37	−39	115	31 58	−39	116	31 19	−40	116	30 39	−40	117	29 59	−40	118
49	33 23	39	115	32 44	40	116	32 04	40	117	31 24	41	118	30 43	40	119
48	34 08	39	116	33 29	40	117	32 49	41	118	32 08	41	119	31 27	41	120
47	34 53	40	117	34 13	40	118	33 33	41	119	32 52	41	120	32 11	42	121
46	35 38	40	118	34 58	41	119	34 17	41	120	33 36	42	121	32 54	42	121
45	36 22	−40	119	35 42	−42	120	35 00	−41	120	34 19	−42	121	33 37	−42	122
44	37 06	41	119	36 25	41	120	35 44	42	121	35 02	43	122	34 19	42	123
43	37 50	42	120	37 08	42	121	36 26	42	122	35 44	43	123	35 01	43	124
42	38 33	42	121	37 51	42	122	37 09	43	123	36 26	43	124	35 43	44	125
41	39 16	42	122	38 34	43	123	37 51	43	124	37 08	44	125	36 24	44	126
40	39 59	−43	123	39 16	−44	124	38 32	−43	125	[illegible]	[illegible]	126	37 05	−45	127
39	40 41	44	124	39 57	43	125	39 14	45	126	38 29	44	127	[illegible]	45	127
38	41 22	44	125	40 38	44	126	39 54	44	127	[illegible]	[illegible]	[illegible]	38 25	46	128
37	42 03	44	126	41 19	45	127	40 34	45	128	39 49	45	129	39 04	46	129
36	42 44	45	127	41 59	45	128	41 14	46	129	40 28	46	130	39 42	46	130
35	43 24	−46	128	42 38	−45	129	41 53	−46	130	41 07	−47	131	40 20	−46	131
34	44 03	46	129	43 17	46	130	42 31	46	131	41 45	47	132	40 58	47	132
33	44 42	46	130	43 56	47	131	43 09	47	132	42 22	47	133	41 35	48	133
32	45 20	47	131	44 33	47	132	43 46	47	133	42 59	48	134	42 11	48	135
31	45 58	47	132	45 11	48	133	44 23	48	134	43 35	48	135	42 47	49	136
30	46 35	−48	133	45 47	−48	134	44 59	−49	135	44 10	−48	136	43 22	−50	137
29	47 11	48	135	46 23	49	135	45 34	49	136	44 45	49	137	43 56	50	138
28	47 47	49	136	46 58	50	137	46 08	49	137	45 19	50	138	44 29	50	139
27	48 21	49	137	47 32	50	138	46 42	50	139	45 52	50	139	45 02	51	140
26	48 55	50	138	48 05	50	139	47 15	51	140	46 24	50	141	45 34	52	141
25	49 28	−50	139	48 38	−51	140	47 47	−51	141	46 56	−51	142	46 05	−52	143
24	50 01	51	141	49 10	52	142	48 18	51	142	47 27	52	143	46 35	53	144
23	50 32	51	142	49 41	52	143	48 49	53	144	47 56	52	144	47 04	53	145
22	51 03	53	143	50 10	52	144	49 18	53	145	48 25	53	146	47 32	53	146
21	51 32	53	145	50 39	53	146	49 46	53	146	48 53	53	147	48 00	54	148
20	52 01	−54	146	51 07	−53	147	50 14	−54	148	49 20	−54	148	48 26	−54	149
19	52 28	54	148	51 34	54	148	50 40	54	149	49 46	55	150	48 51	54	150
18	52 54	54	149	52 00	55	150	51 05	54	151	50 11	55	151	49 16	55	152
17	53 19	54	151	52 25	55	151	51 30	56	152	50 34	55	153	49 39	56	153
16	53 44	56	152	52 48	55	153	51 53	56	154	50 57	56	154	50 01	56	155
15	54 06	−55	154	53 11	−56	154	52 15	−57	155	51 18	−56	156	50 22	−57	156
14	54 28	56	155	53 32	57	156	52 35	56	157	51 39	57	157	50 42	57	158
13	54 48	57	157	53 51	56	158	52 55	58	158	51 57	57	159	51 00	57	159
12	55 07	57	159	54 10	57	159	53 13	58	160	52 15	57	160	51 18	58	161
11	55 25	58	160	54 27	58	161	53 29	57	161	52 32	58	162	51 34	58	162
10	55 41	−58	162	54 43	−58	163	53 45	−58	163	52 47	−59	163	51 48	−58	164
9	55 56	59	164	54 57	58	164	53 59	59	165	53 00	58	165	52 02	59	165
8	56 09	59	166	55 10	58	166	54 12	59	166	53 13	59	167	52 14	59	167
7	56 21	59	167	55 22	59	168	54 23	59	168	53 24	59	168	52 25	60	169
6	56 31	59	169	55 32	59	169	54 33	60	170	53 33	59	170	52 34	59	170
5	56 40	−59	171	55 41	−60	171	54 41	−60	171	53 41	−59	172	52 42	−60	172
4	56 47	59	173	55 48	60	173	54 48	60	173	53 48	60	173	52 48	59	173
3	56 53	60	175	55 53	60	175	54 53	60	175	53 53	59	175	52 54	60	175
2	56 57	60	176	55 57	60	176	54 57	60	177	53 57	60	177	52 57	60	177
1	56 59	60	178	55 59	60	178	54 59	60	178	53 59	60	178	52 59	60	178
0	[illegible]	[illegible]	[illegible]	[illegible]	[illegible]	[illegible]	[illegible]	[illegible]	[illegible]	[illegible]	[illegible]	[illegible]	[illegible]	[illegible]	[illegible]

LHA	5° Hc	d	Z	6° Hc	d	Z	7° Hc	d	Z	8° Hc	d	Z	9° Hc	d	Z
	° ′	′	°	° ′	′	°	° ′	′	°	° ′	′	°	° ′	′	°
69	14 36	36	106	14 00	35	107	13 25	36	108	12 49	36	109	12 13	36	109
68	15 24	36	107	14 48	35	108	14 13	36	108	13 37	36	109	13 01	37	110
67	16 12	36	107	15 36	36	108	15 00	36	109	14 24	36	110	13 48	37	111
66	17 00	36	108	16 24	36	109	15 48	37	110	15 11	36	110	14 35	37	111
65	17 48	−36	109	17 12	−37	109	16 35	−37	110	15 58	−36	111	15 22	−37	112
64	18 35	36	109	17 59	37	110	17 22	37	111	16 45	37	112	16 08	37	113
63	19 23	37	110	18 46	37	111	18 09	37	112	17 32	37	112	16 55	38	113
62	20 10	37	110	19 33	37	111	18 56	38	112	18 18	37	113	17 41	38	114
61	20 57	37	111	20 20	38	112	19 42	37	113	19 05	38	113	18 27	38	114
60	21 44	−38	112	21 06	−37	113	20 29	−38	113	19 51	−39	114	19 12	−38	115
59	22 31	38	112	21 53	38	113	21 15	39	114	20 36	38	115	19 58	39	116
58	23 17	38	113	22 39	38	114	22 01	39	115	21 22	39	116	20 43	39	116
57	24 03	38	114	23 25	39	115	22 46	39	116	22 07	39	116	21 28	39	117
56	24 49	39	115	24 10	39	115	23 31	39	116	22 52	39	117	22 13	40	118
55	25 35	−39	115	24 56	−40	116	24 16	−39	117	23 37	−40	118	22 57	−40	119
54	26 20	39	116	25 41	40	117	25 01	40	118	24 21	40	118	23 41	40	119
53	27 05	39	117	26 26	40	117	25 46	41	118	25 05	40	119	24 25	41	120
52	27 50	40	117	27 10	40	118	26 30	41	119	25 49	41	120	25 08	41	121
51	28 35	41	118	27 54	40	119	27 14	41	120	26 33	42	121	25 51	41	122
50	29 19	−41	119	28 38	−41	120	27 57	−41	121	27 16	−42	121	26 34	−42	122
49	30 03	41	120	29 22	42	121	28 40	42	121	27 58	42	122	27 16	42	123
48	30 46	41	121	30 05	42	121	29 23	42	122	28 41	43	123	27 58	42	124
47	31 29	41	121	30 48	43	122	30 05	42	123	29 23	43	124	28 40	43	125
46	32 12	42	122	31 30	43	123	30 47	43	124	30 04	43	125	29 21	43	125
45	32 55	−43	123	32 12	−43	124	31 29	−43	125	30 46	−44	125	30 02	−44	126
44	33 37	43	124	32 54	44	125	32 10	43	126	31 27	44	126	30 43	45	127
43	34 18	43	125	33 35	44	126	32 51	44	126	32 07	45	127	31 22	44	128
42	34 59	44	126	34 15	44	126	33 31	44	127	32 47	45	128	32 02	45	129
41	35 40	44	126	34 56	45	127	34 11	45	128	33 26	45	129	32 41	46	130
40	36 20	−44	127	35 36	−46	128	34 50	−45	129	34 05	−46	130	33 19	−46	131
39	37 00	45	128	36 15	46	129	35 29	45	130	34 44	47	131	33 57	46	132
38	37 39	45	129	36 54	46	130	36 08	47	131	35 21	46	132	34 35	47	132
37	38 18	46	130	37 32	47	131	36 45	46	132	35 59	47	133	35 12	48	133
36	38 56	46	131	38 10	47	132	37 23	47	133	36 36	48	134	35 48	48	134
35	39 34	−47	132	38 47	−48	133	37 59	−47	134	37 12	−48	135	36 24	−49	135
34	40 11	48	133	39 23	48	134	38 35	48	135	37 47	48	136	36 59	49	136
33	40 47	48	134	39 59	48	135	39 11	49	136	38 22	49	137	37 33	49	137
32	41 23	49	135	40 34	48	136	39 46	50	137	38 56	49	138	38 07	50	138
31	41 58	49	136	41 09	49	137	40 20	50	138	39 30	50	139	38 40	50	139
30	42 32	−49	138	41 43	−50	138	40 53	−50	139	40 03	−50	140	39 13	−51	140
29	43 06	50	139	42 16	50	139	41 26	51	140	40 35	51	141	39 44	51	142
28	43 39	51	140	42 48	50	141	41 58	51	141	41 07	52	142	40 15	51	143
27	44 11	51	141	43 20	51	142	42 29	52	142	41 37	51	143	40 46	52	144
26	44 42	51	142	43 51	52	143	42 59	52	144	42 07	52	144	41 15	53	145
25	45 13	−52	143	44 21	−52	144	43 29	−53	145	42 36	−53	145	41 43	−52	146
24	45 42	52	145	44 50	53	145	43 57	53	146	43 04	53	147	42 11	53	147
23	46 11	53	146	45 18	53	147	44 25	53	147	43 32	54	148	42 38	54	148
22	46 39	53	147	45 46	54	148	44 52	54	148	43 58	54	149	43 04	54	150
21	47 06	54	148	46 12	54	149	45 18	55	150	44 23	54	150	43 29	55	151
20	47 32	−55	150	46 37	−54	150	45 43	−55	151	44 48	−55	152	43 53	−55	152
19	47 57	55	151	47 02	55	152	46 07	55	152	45 12	56	153	44 16	55	153
18	48 21	56	152	47 25	55	153	46 30	56	154	45 34	56	154	44 38	56	155
17	48 43	55	154	47 48	56	154	46 52	56	155	45 56	57	155	44 59	56	156
16	49 05	56	155	48 09	57	156	47 12	56	156	46 16	57	157	45 19	56	157
15	49 25	−56	157	48 29	−57	157	47 32	−57	158	46 35	−57	158	45 38	−57	159
14	49 45	57	158	48 48	57	159	47 51	58	159	46 53	57	160	45 56	57	160
13	50 03	57	160	49 06	58	160	48 08	57	161	47 11	58	161	46 13	58	161
12	50 20	58	161	49 22	58	162	48 24	58	162	47 26	58	162	46 28	58	163
11	50 36	58	163	49 38	59	163	48 39	58	163	47 41	58	164	46 43	59	164
10	50 50	−58	164	49 52	−59	165	48 53	−58	165	47 55	−59	165	46 56	−59	166
9	51 03	58	166	50 05	59	166	49 06	59	166	48 07	59	167	47 08	59	167
8	51 15	59	167	50 16	59	168	49 17	59	168	48 18	59	168	47 19	59	168
7	51 25	59	169	50 26	59	169	49 27	59	169	48 28	60	170	47 28	59	170
6	51 35	60	170	50 35	59	171	49 36	60	171	48 36	59	171	47 37	60	171
5	51 42	−59	172	50 43	−60	172	49 43	−59	172	48 44	−60	173	47 44	−60	173
4	51 49	60	174	50 49	60	174	49 49	60	174	48 49	59	174	47 50	60	174
3	51 54	60	175	50 54	60	175	49 54	60	175	48 54	60	176	47 54	60	176
2	51 57	60	177	50 57	60	177	49 57	60	177	48 57	60	177	47 57	59	177
1	51 59	60	178	50 59	60	178	49 59	60	179	48 59	60	179	47 59	60	179
0	52 00	[illegible]	180	51 00	−60	180	50 00	−60	180	49 00	−60	180	48 00	−60	180

LHA	10° Hc	d	Z	11° Hc	d	Z	12° Hc	d	Z	13° Hc	d	Z	14° Hc	d	Z	LHA
	° ′	′	°	° ′	′	°	° ′	′	°	° ′	′	°	° ′	′	°	
69	11 37	36	110	11 01	36	111	10 25	37	112	09 48	36	113	09 12	37	113	291
68	12 24	36	111	11 48	37	112	11 11	36	112	10 35	37	113	09 58	37	114	292
67	13 11	36	111	12 35	37	112	11 58	37	113	11 21	37	114	10 44	37	115	293
66	13 58	37	112	13 21	37	113	12 44	37	114	12 07	37	114	11 30	38	115	294
65	14 45	−38	113	14 07	−37	114	13 30	−37	114	12 53	−38	115	12 15	−38	116	295
64	15 31	38	113	14 53	37	114	14 16	38	115	13 38	38	116	13 00	38	117	296
63	16 17	38	114	15 39	38	115	15 01	38	116	14 23	38	116	13 45	38	117	297
62	17 03	38	115	16 25	38	115	15 47	39	116	15 08	38	117	14 30	39	118	298
61	17 49	39	115	17 10	38	116	16 32	39	117	15 53	39	118	15 14	39	118	299
60	18 34	−39	116	17 55	−39	117	17 16	−39	118	16 37	−39	118	15 58	−39	119	300
59	19 19	39	117	18 40	39	117	18 01	39	118	17 22	40	119	16 42	40	120	301
58	20 04	39	117	19 25	40	118	18 45	40	119	18 05	39	120	17 26	41	120	302
57	20 49	40	118	20 09	40	119	19 29	40	120	18 49	40	120	18 09	41	121	303
56	21 33	40	119	20 53	40	119	20 13	41	120	19 32	40	121	18 52	41	122	304
55	22 17	−40	119	21 37	−41	120	20 56	−41	121	20 15	−41	122	19 34	−41	123	305
54	23 01	41	120	22 20	41	121	21 39	41	122	20 58	41	122	20 17	42	123	306
53	23 44	41	121	23 03	41	122	22 22	42	122	21 40	41	123	20 59	42	124	307
52	24 27	41	122	23 46	42	122	23 04	42	123	22 22	42	124	21 40	42	125	308
51	25 10	42	122	24 28	42	123	23 46	42	124	23 04	43	125	22 21	42	125	309
50	25 52	−42	123	25 10	−42	124	24 28	−43	125	23 45	−43	125	23 02	−43	126	310
49	26 34	42	124	25 52	43	125	25 09	43	125	24 26	43	126	23 43	44	127	311
48	27 16	43	125	26 33	43	125	25 50	44	126	25 06	43	127	24 23	44	128	312
47	27 57	43	125	27 14	44	126	26 30	44	127	25 46	44	128	25 02	44	128	313
46	28 38	44	126	27 54	44	127	27 10	44	128	26 26	44	129	25 42	45	129	314
45	29 18	−44	127	28 34	−44	128	27 50	−45	129	27 05	−45	129	26 20	−45	130	315
44	29 58	44	128	29 14	45	129	28 29	45	129	27 44	45	130	26 59	46	131	316
43	30 38	45	129	29 53	45	130	29 08	46	130	28 22	46	131	27 36	46	132	317
42	31 17	46	130	30 31	45	130	29 46	46	131	29 00	46	132	28 14	47	133	318
41	31 55	45	130	31 10	47	131	30 23	46	132	29 37	46	133	28 51	47	133	319
40	32 33	−46	131	31 47	−46	132	31 01	−47	133	30 14	−47	134	29 27	−47	134	320
39	33 11	47	132	32 24	47	133	31 37	47	134	30 50	47	134	30 03	48	135	321
38	33 48	47	133	33 01	48	134	32 13	47	135	31 26	48	135	30 38	48	136	322
37	34 24	47	134	33 37	48	135	32 49	48	136	32 01	49	136	31 12	48	137	323
36	35 00	48	135	34 12	48	136	33 24	49	137	32 35	48	137	31 47	49	138	324
35	35 35	−48	136	34 47	−49	137	33 58	−49	137	33 09	−49	138	32 20	−49	139	325
34	36 10	49	137	35 21	49	138	34 32	50	138	33 42	49	139	32 53	50	140	326
33	36 44	49	138	35 55	50	139	35 05	50	139	34 15	50	140	33 25	50	141	327
32	37 17	49	139	36 28	51	140	35 37	50	140	34 47	50	141	33 57	51	142	328
31	37 50	50	140	37 00	51	141	36 09	51	141	35 18	51	142	34 27	51	143	329
30	38 22	−51	141	37 31	−51	142	36 40	−51	142	35 49	−51	143	34 58	−52	144	330
29	38 53	51	142	38 02	51	143	37 11	52	144	36 19	52	144	35 27	52	145	331
28	39 24	52	143	38 32	52	144	37 40	52	145	36 48	52	145	35 56	53	146	332
27	39 54	53	144	39 01	52	145	38 09	53	146	37 16	52	146	36 24	53	147	333
26	40 22	52	146	39 30	53	146	38 37	53	147	37 44	53	147	36 51	54	148	334
25	40 51	−54	147	39 57	−53	147	39 04	−53	148	38 11	−54	148	37 17	−54	149	335
24	41 18	54	148	40 24	53	148	39 31	54	149	38 37	54	150	37 43	55	150	336
23	41 44	54	149	40 50	54	150	39 56	54	150	39 02	55	151	38 07	54	151	337
22	42 10	55	150	41 15	54	151	40 21	55	151	39 26	55	152	38 31	55	152	338
21	42 34	55	151	41 39	55	152	40 44	55	152	39 49	55	153	38 54	55	154	339
20	42 58	−55	153	42 03	−56	153	41 07	−55	154	40 12	−56	154	39 16	−56	155	340
19	43 21	56	154	42 25	56	154	41 29	56	155	40 33	56	155	39 37	56	156	341
18	43 42	56	155	42 46	56	156	41 50	56	156	40 54	57	157	39 57	56	157	342
17	44 03	57	156	43 06	56	157	42 10	57	157	41 13	57	158	40 16	56	158	343
16	44 23	58	158	43 26	57	158	42 29	57	159	41 32	57	159	40 35	57	159	344
15	44 41	−57	159	43 44	−57	159	42 47	−58	160	41 49	−57	160	40 52	−58	161	345
14	44 59	58	160	44 01	57	161	43 04	58	161	42 06	58	162	41 08	58	162	346
13	45 15	58	162	44 17	58	162	43 19	58	163	42 21	58	163	41 23	58	163	347
12	45 30	58	163	44 32	58	163	43 34	58	164	42 36	59	164	41 37	58	164	348
11	45 44	58	164	44 46	58	165	43 48	59	165	42 49	59	165	41 50	58	166	349
10	45 57	−58	166	44 59	−59	166	44 00	−59	166	43 01	−59	167	42 02	−58	167	350
9	46 09	59	167	45 10	59	167	44 11	59	168	43 12	59	168	42 13	59	168	351
8	46 20	59	169	45 21	60	169	44 21	59	169	43 22	59	169	42 23	59	170	352
7	46 29	59	170	45 30	60	170	44 30	59	170	43 31	59	171	42 32	60	171	353
6	46 37	59	171	45 38	60	172	44 38	60	172	43 39	60	172	42 39	59	172	354
5	46 44	−59	173	45 45	−60	173	44 45	−60	173	43 45	−59	173	42 46	−60	173	355
4	46 50	60	174	45 50	60	174	44 50	59	175	43 51	60	175	42 51	60	175	356
3	46 54	60	176	45 54	59	176	44 55	60	176	43 55	60	176	42 55	60	176	357
2	46 58	60	177	45 58	60	177	44 58	60	177	43 58	60	177	42 58	60	177	358
1	46 59	60	179	45 59	60	179	44 59	60	179	43 59	60	179	42 59	60	179	359
0	47 00	−60	180	46 00	−60	180	45 00	−60	180	44 00	−60	180	43 00	−60	180	360

LAT 33°

200

TABLE V (Addendum 6) HO249 Vol II page 20
Lat 33° Declination 0°-14° Contrar

TABLE 5.—Correction to Tabulated Altitude for Minutes of Declination

d / ′	1	2	3	4	5	6	7	8	9	10	11	12	13	14	15	16	17	18	19	20	21	22	23	24	25	26	27	28	29	30	31	32	33	34	35	36	37	38	39	40	41	42	43	44	45	46	47	48	49	50	51	52	53	54	55	56	57	58	59	60	d / ′
0	0	0	0	0	0	0	0	0	0	0	0	0	0	0	0	0	0	0	0	0	0	0	0	0	0	0	0	0	0	0	0	0	0	0	0	0	0	0	0	0	0	0	0	0	0	0	0	0	0	0	0	0	0	0	0	0	0	0	0	0	0
1	0	0	0	0	0	0	0	0	0	0	0	0	0	0	0	0	0	0	0	0	0	0	0	0	0	0	0	0	0	0	1	1	1	1	1	1	1	1	1	1	1	1	1	1	1	1	1	1	1	1	1	1	1	1	1	1	1	1	1	1	1
2	0	0	0	0	0	0	0	0	0	0	0	0	0	0	0	1	1	1	1	1	1	1	1	1	1	1	1	1	1	1	1	1	1	1	1	1	1	1	1	1	1	1	1	1	2	2	2	2	2	2	2	2	2	2	2	2	2	2	2	2	2
3	0	0	0	0	0	0	0	0	0	0	1	1	1	1	1	1	1	1	1	1	1	1	1	1	1	1	1	1	1	2	2	2	2	2	2	2	2	2	2	2	2	2	2	2	2	2	2	2	2	2	3	3	3	3	3	3	3	3	3	3	3
4	0	0	0	0	0	0	0	1	1	1	1	1	1	1	1	1	1	1	1	1	1	1	2	2	2	2	2	2	2	2	2	2	2	2	2	2	2	3	3	3	3	3	3	3	3	3	3	3	3	3	3	3	4	4	4	4	4	4	4	4	4
5	0	0	0	0	0	0	1	1	1	1	1	1	1	1	1	1	1	2	2	2	2	2	2	2	2	2	2	2	2	2	3	3	3	3	3	3	3	3	3	3	3	4	4	4	4	4	4	4	4	4	4	4	4	4	5	5	5	5	5	5	5
6	0	0	0	0	0	1	1	1	1	1	1	1	1	1	2	2	2	2	2	2	2	2	2	2	2	3	3	3	3	3	3	3	3	3	4	4	4	4	4	4	4	4	4	4	4	5	5	5	5	5	5	5	5	5	6	6	6	6	6	6	6
7	0	0	0	0	1	1	1	1	1	1	1	1	2	2	2	2	2	2	2	2	2	3	3	3	3	3	3	3	3	4	4	4	4	4	4	4	4	4	5	5	5	5	5	5	5	5	5	6	6	6	6	6	6	6	6	7	7	7	7	7	7
8	0	0	0	1	1	1	1	1	1	1	1	2	2	2	2	2	2	2	3	3	3	3	3	3	3	3	4	4	4	4	4	4	4	5	5	5	5	5	5	5	5	6	6	6	6	6	6	6	7	7	7	7	7	7	7	7	8	8	8	8	8
9	0	0	0	1	1	1	1	1	1	2	2	2	2	2	2	2	3	3	3	3	3	3	3	4	4	4	4	4	4	4	5	5	5	5	5	5	6	6	6	6	6	6	6	7	7	7	7	7	7	8	8	8	8	8	8	8	9	9	9	9	9
10	0	0	0	1	1	1	1	1	2	2	2	2	2	2	2	3	3	3	3	3	4	4	4	4	4	4	4	5	5	5	5	5	6	6	6	6	6	6	6	7	7	7	7	7	8	8	8	8	8	8	8	9	9	9	9	9	10	10	10	10	10
11	0	0	1	1	1	1	1	1	2	2	2	2	2	3	3	3	3	3	3	4	4	4	4	4	5	5	5	5	5	6	6	6	6	6	6	7	7	7	7	7	8	8	8	8	8	8	9	9	9	9	9	10	10	10	10	10	10	11	11	11	11
12	0	0	1	1	1	1	1	2	2	2	2	2	3	3	3	3	3	4	4	4	4	4	5	5	5	5	5	6	6	6	6	6	7	7	7	7	7	8	8	8	8	8	9	9	9	9	9	10	10	10	10	10	11	11	11	11	11	12	12	12	12
13	0	0	1	1	1	1	2	2	2	2	2	3	3	3	3	3	4	4	4	4	5	5	5	5	5	6	6	6	6	6	7	7	7	7	8	8	8	8	8	9	9	9	9	10	10	10	10	10	11	11	11	11	11	12	12	12	12	13	13	13	13
14	0	0	1	1	1	1	2	2	2	2	3	3	3	3	4	4	4	4	4	5	5	5	5	6	6	6	6	7	7	7	7	7	8	8	8	8	9	9	9	9	10	10	10	10	10	11	11	11	11	12	12	12	12	13	13	13	13	14	14	14	14
15	0	0	1	1	1	2	2	2	2	2	3	3	3	4	4	4	4	4	5	5	5	6	6	6	6	6	7	7	7	8	8	8	8	8	9	9	9	10	10	10	10	10	11	11	11	12	12	12	12	12	13	13	13	14	14	14	14	14	15	15	15
16	0	1	1	1	1	2	2	2	2	3	3	3	3	4	4	4	5	5	5	5	6	6	6	6	7	7	7	7	8	8	8	9	9	9	9	10	10	10	10	11	11	11	11	12	12	12	13	13	13	13	14	14	14	14	15	15	15	15	16	16	16
17	0	1	1	1	1	2	2	2	3	3	3	3	4	4	4	5	5	5	5	6	6	6	7	7	7	7	8	8	8	8	9	9	9	10	10	10	10	11	11	11	12	12	12	12	13	13	13	14	14	14	14	15	15	15	16	16	16	16	17	17	17
18	0	1	1	1	2	2	2	2	3	3	3	4	4	4	4	5	5	5	6	6	6	7	7	7	8	8	8	8	9	9	9	10	10	10	10	11	11	11	12	12	12	13	13	13	14	14	14	14	15	15	15	16	16	16	16	17	17	17	18	18	18
19	0	1	1	1	2	2	2	3	3	3	3	4	4	4	5	5	5	6	6	6	7	7	7	8	8	8	9	9	9	10	10	10	10	11	11	11	12	12	12	13	13	13	14	14	14	15	15	15	16	16	16	16	17	17	17	18	18	18	19	19	19
20	0	1	1	1	2	2	2	3	3	3	4	4	4	5	5	5	6	6	6	7	7	7	8	8	8	9	9	9	10	10	10	11	11	11	12	12	12	13	13	13	14	14	14	15	15	15	16	16	16	17	17	17	18	18	18	19	19	19	20	20	20
21	0	1	1	1	2	2	2	3	3	4	4	4	5	5	5	6	6	6	7	7	7	8	8	8	9	9	9	10	10	10	11	11	12	12	12	13	13	13	14	14	14	15	15	15	16	16	16	17	17	18	18	18	19	19	19	20	20	20	21	21	21
22	0	1	1	1	2	2	3	3	3	4	4	4	5	5	6	6	6	7	7	7	8	8	8	9	9	10	10	10	11	11	11	12	12	12	13	13	14	14	14	15	15	15	16	16	16	17	17	18	18	18	19	19	19	20	20	21	21	21	22	22	22
23	0	1	1	2	2	2	3	3	3	4	4	5	5	5	6	6	7	7	7	8	8	8	9	9	10	10	10	11	11	12	12	12	13	13	13	14	14	15	15	15	16	16	16	17	17	18	18	18	19	19	20	20	20	21	21	21	22	22	23	23	23
24	0	1	1	2	2	2	3	3	4	4	4	5	5	6	6	6	7	7	8	8	8	9	9	10	10	10	11	11	12	12	12	13	13	14	14	14	15	15	16	16	16	17	17	18	18	18	19	19	20	20	20	21	21	22	22	22	23	23	24	24	24
25	0	1	1	2	2	2	3	3	4	4	5	5	5	6	6	7	7	8	8	8	9	9	10	10	10	11	11	12	12	12	13	13	14	14	15	15	15	16	16	17	17	18	18	18	19	19	20	20	20	21	21	22	22	22	23	23	24	24	25	25	25
26	0	1	1	2	2	3	3	3	4	4	5	5	6	6	6	7	7	8	8	9	9	10	10	10	11	11	12	12	13	13	13	14	14	15	15	16	16	16	17	17	18	18	19	19	20	20	20	21	21	22	22	23	23	23	24	24	25	25	26	26	26
27	0	1	1	2	2	3	3	4	4	4	5	5	6	6	7	7	8	8	9	9	9	10	10	11	11	12	12	13	13	14	14	14	15	15	16	16	17	17	18	18	18	19	19	20	20	21	21	22	22	22	23	23	24	24	25	25	26	26	27	27	27
28	0	1	1	2	2	3	3	4	4	5	5	6	6	7	7	7	8	8	9	9	10	10	11	11	12	12	13	13	14	14	14	15	15	16	16	17	17	18	18	19	19	20	20	21	21	21	22	22	23	23	24	24	25	25	26	26	27	27	28	28	28
29	0	1	1	2	2	3	3	4	4	5	5	6	6	7	7	8	8	9	9	10	10	11	11	12	12	13	13	14	14	14	15	15	16	16	17	17	18	18	19	19	20	20	21	21	22	22	23	23	24	24	25	25	26	26	27	27	28	28	29	29	29
30	0	1	2	2	2	3	4	4	4	5	6	6	6	7	8	8	8	9	10	10	10	11	12	12	12	13	14	14	14	15	16	16	16	17	18	18	18	19	20	20	20	21	22	22	22	23	24	24	24	25	26	26	26	27	28	28	28	29	30	30	30
31	1	1	2	2	3	3	4	4	5	5	6	6	7	7	8	8	9	9	10	10	11	11	12	12	13	13	14	14	15	16	16	17	17	18	18	19	19	20	20	21	21	22	22	23	23	24	24	25	25	26	26	27	27	28	28	29	29	30	30	31	31
32	1	1	2	2	3	3	4	4	5	5	6	6	7	7	8	9	9	10	10	11	11	12	12	13	13	14	14	15	15	16	17	17	18	18	19	19	20	20	21	21	22	22	23	23	24	25	25	26	26	27	27	28	28	29	29	30	30	31	31	32	32
33	1	1	2	2	3	3	4	4	5	6	6	7	7	8	8	9	9	10	10	11	12	12	13	13	14	14	15	15	16	16	17	18	18	19	19	20	20	21	21	22	23	23	24	24	25	25	26	26	27	28	28	29	29	30	30	31	31	32	32	33	33
34	1	1	2	2	3	3	4	5	5	6	6	7	7	8	8	9	10	10	11	11	12	12	13	14	14	15	15	16	16	17	18	18	19	19	20	20	21	22	22	23	23	24	24	25	26	26	27	27	28	28	29	29	30	31	31	32	32	33	33	34	34
35	1	1	2	2	3	4	4	5	5	6	6	7	8	8	9	9	10	10	11	12	12	13	13	14	15	15	16	16	17	18	18	19	19	20	20	21	22	22	23	23	24	24	25	26	26	27	27	28	29	29	30	30	31	32	32	33	33	34	34	35	35
36	1	1	2	2	3	4	4	5	5	6	7	7	8	8	9	10	10	11	11	12	13	13	14	14	15	16	16	17	17	18	19	19	20	20	21	22	22	23	23	24	25	25	26	26	27	28	28	29	29	30	31	31	32	32	33	34	34	35	35	36	36
37	1	1	2	2	3	4	4	5	6	6	7	7	8	9	9	10	10	11	12	12	13	14	14	15	15	16	17	17	18	18	19	20	20	21	22	22	23	23	24	25	25	26	27	27	28	28	29	30	30	31	31	32	33	33	34	35	35	36	36	37	37
38	1	1	2	3	3	4	4	5	6	6	7	8	8	9	10	10	11	11	12	13	13	14	15	15	16	16	17	18	18	19	20	20	21	22	22	23	23	24	25	25	26	27	27	28	28	29	30	30	31	32	32	33	34	34	35	35	36	37	37	38	38
39	1	1	2	3	3	4	5	5	6	6	7	8	8	9	10	10	11	12	12	13	14	14	15	16	16	17	18	18	19	20	20	21	21	22	23	23	24	25	25	26	27	27	28	29	29	30	31	31	32	32	33	34	34	35	36	36	37	38	38	39	39
40	1	1	2	3	3	4	5	5	6	7	7	8	9	9	10	11	11	12	13	13	14	15	15	16	17	17	18	19	19	20	21	21	22	23	23	24	25	25	26	27	27	28	29	29	30	31	31	32	33	33	34	35	35	36	37	37	38	39	39	40	40
41	1	1	2	3	3	4	5	5	6	7	8	8	9	10	10	11	12	12	13	14	14	15	16	16	17	18	18	19	20	20	21	22	23	23	24	25	25	26	27	27	28	29	29	30	31	31	32	33	33	34	35	36	36	37	38	38	39	40	40	41	41
42	1	1	2	3	4	4	5	6	6	7	8	8	9	10	10	11	12	13	13	14	15	15	16	17	18	18	19	20	20	21	22	22	23	24	24	25	26	27	27	28	29	29	30	31	32	32	33	34	34	35	36	36	37	38	38	39	40	41	41	42	42
43	1	1	2	3	4	4	5	6	6	7	8	9	9	10	11	11	12	13	14	14	15	16	16	17	18	19	19	20	21	22	22	23	24	24	25	26	27	27	28	29	29	30	31	32	32	33	34	34	35	36	37	37	38	39	39	40	41	42	42	43	43
44	1	1	2	3	4	4	5	6	7	7	8	9	10	10	11	12	12	13	14	15	15	16	17	18	18	19	20	21	21	22	23	23	24	25	26	26	27	28	29	29	30	31	32	32	33	34	34	35	36	37	37	38	39	40	40	41	42	43	43	44	44
45	1	2	2	3	4	4	5	6	7	8	8	9	10	10	11	12	13	14	14	15	16	16	17	18	19	20	20	21	22	22	23	24	25	26	26	27	28	28	29	30	31	32	32	33	34	34	35	36	37	38	38	39	40	40	41	42	43	44	44	45	45
46	1	2	2	3	4	5	5	6	7	8	8	9	10	11	12	12	13	14	15	15	16	17	18	18	19	20	21	21	22	23	24	25	25	26	27	28	28	29	30	31	31	32	33	34	34	35	36	37	38	38	39	40	41	41	42	43	44	44	45	46	46
47	1	2	2	3	4	5	5	6	7	8	9	9	10	11	12	13	13	14	15	16	16	17	18	19	20	20	21	22	23	24	24	25	26	27	27	28	29	30	31	31	32	33	34	34	35	36	37	38	38	39	40	41	42	42	43	44	45	45	46	47	47
48	1	2	2	3	4	5	6	6	7	8	9	10	10	11	12	13	14	14	15	16	17	18	18	19	20	21	22	22	23	24	25	26	26	27	28	29	30	30	31	32	33	34	34	35	36	37	38	38	39	40	41	42	42	43	44	45	46	46	47	48	48
49	1	2	2	3	4	5	6	7	7	8	9	10	11	11	12	13	14	15	16	16	17	18	19	20	20	21	22	23	24	24	25	26	27	28	29	29	30	31	32	33	33	34	35	36	37	38	38	39	40	41	42	42	43	44	45	46	47	47	48	49	49
50	1	2	2	3	4	5	6	7	8	8	9	10	11	12	12	13	14	15	16	17	18	18	19	20	21	22	22	23	24	25	26	27	28	28	29	30	31	32	32	33	34	35	36	37	38	38	39	40	41	42	42	43	44	45	46	47	48	48	49	50	50
51	1	2	3	3	4	5	6	7	8	8	9	10	11	12	13	14	14	15	16	17	18	19	20	20	21	22	23	24	25	26	26	27	28	29	30	31	31	32	33	34	35	36	37	37	38	39	40	41	42	42	43	44	45	46	47	48	48	49	50	51	51
52	1	2	3	3	4	5	6	7	8	9	10	10	11	12	13	14	15	16	16	17	18	19	20	21	22	23	23	24	25	26	27	28	29	29	30	31	32	33	34	35	36	36	37	38	39	40	41	42	42	43	44	45	46	47	48	49	49	50	51	52	52
53	1	2	3	4	4	5	6	7	8	9	10	11	11	12	13	14	15	16	17	18	19	19	20	21	22	23	24	25	26	26	27	28	29	30	31	32	33	34	34	35	36	37	38	39	40	41	42	42	43	44	45	46	47	48	49	49	50	51	52	53	53
54	1	2	3	4	4	5	6	7	8	9	10	11	12	13	14	14	15	16	17	18	19	20	21	22	22	23	24	25	26	27	28	29	30	31	32	32	33	34	35	36	37	38	39	40	40	41	42	43	44	45	46	47	48	49	50	50	51	52	53	54	54
55	1	2	3	4	5	6	6	7	8	9	10	11	12	13	14	15	16	16	17	18	19	20	21	22	23	24	25	26	27	28	28	29	30	31	32	33	34	35	36	37	38	38	39	40	41	42	43	44	45	46	47	48	49	50	50	51	52	53	54	55	55
56	1	2	3	4	5	6	7	7	8	9	10	11	12	13	14	15	16	17	18	19	20	21	21	22	23	24	25	26	27	28	29	30	31	32	33	34	35	35	36	37	38	39	40	41	42	43	44	45	46	47	48	49	49	50	51	52	53	54	55	56	56
57	1	2	3	4	5	6	7	8	9	10	10	11	12	13	14	15	16	17	18	19	20	21	22	23	24	25	26	27	28	28	29	30	31	32	33	34	35	36	37	38	39	40	41	42	43	44	45	46	47	48	48	49	50	51	52	53	54	55	56	57	57
58	1	2	3	4	5	6	7	8	9	10	11	12	13	14	14	15	16	17	18	19	20	21	22	23	24	25	26	27	28	29	30	31	32	33	34	35	36	37	38	39	40	41	42	43	44	44	45	46	47	48	49	50	51	52	53	54	55	56	57	58	58
59	1	2	3	4	5	6	7	8	9	10	11	12	13	14	15	16	17	18	19	20	21	22	23	24	25	26	27	28	29	30	30	31	32	33	34	35	36	37	38	39	40	41	42	43	44	45	46	47	48	49	50	51	52	53	54	55	56	57	58	59	59

242

TABLE VI (Addendum 8) HO249 Vol II page 242

Table 5 - Correction to Tabulated Altitude

GMT	☉ SUN GHA	SUN Dec.	ARIES GHA ♈	VENUS –3.7 GHA	VENUS Dec.	MARS –1.9 GHA	MARS Dec.	JUPITER –2.1 GHA	JUPITER Dec.	◐ MOON GHA	MOON Dec.
h m	° ′	° ′	° ′	° ′	° ′	° ′	° ′	° ′	° ′	° ′	° ′
00 00	182 37.1	S 3 26.0	10 34.2	141 52	S19 53	333 45	N11 23	65 48	S20 24	120 08	S23 59
10	185 07.2	26.1	13 04.6	144 22		336 16		68 19		122 33	23 59
20	187 37.2	26.3	15 35.1	146 52		338 46		70 49		124 58	24 00
30	190 07.2	· 26.5	18 05.5	149 21	·	341 17	·	73 20	·	127 23	· 00
40	192 37.3	26.6	20 35.9	151 51		343 47		75 50		129 48	00
50	195 07.3	26.8	23 06.3	154 21		346 18		78 20		132 13	00
01 00	197 37.3	S 3 27.0	25 36.7	156 51	S19 54	348 48	N11 23	80 51	S20 24	134 38	S24 01
10	200 07.4	27.1	28 07.1	159 21		351 19		83 21		137 03	01
20	202 37.4	27.3	30 37.5	161 51		353 49		85 52		139 28	01
30	205 07.4	· 27.4	33 07.9	164 21	·	356 20	·	88 22	·	141 53	· 01
40	207 37.5	27.6	35 38.3	166 51		358 50		90 52		144 18	02
50	210 07.5	27.8	38 08.7	169 21		1 21		93 23		146 43	02
02 00	212 37.5	S 3 27.9	40 39.2	171 51	S19 55	3 51	N11 23	95 53	S20 24	149 07	S24 02
10	215 07.6	28.1	43 09.6	174 21		6 21		98 24		151 32	02
20	217 37.6	28.2	45 40.0	176 51		8 52		100 54		153 57	03
30	220 07.6	· 28.4	48 10.4	179 21	·	11 22	·	103 24	·	156 22	· 03
40	222 37.7	28.6	50 40.8	181 51		13 53		105 55		158 47	03
50	225 07.7	28.7	53 11.2	184 20		16 23		108 25		161 12	03
03 00	227 37.7	S 3 28.9	55 41.6	186 50	S19 56	18 54	N11 22	110 56	S20 24	163 37	S24 03
10	230 07.8	29.1	58 12.0	189 20		21 24		113 26		166 02	03
20	232 37.8	29.2	60 42.4	191 50		23 55		115 56		168 27	04
30	235 07.8	· 29.4	63 12.9	194 20	·	26 25	·	118 27	·	170 52	· 04
40	237 37.9	29.5	65 43.3	196 50		28 56		120 57		173 17	04
50	240 07.9	29.7	68 13.7	199 20		31 26		123 28		175 42	04
04 00	242 37.9	S 3 29.9	70 44.1	201 50	S19 57	33 57	N11 22	125 58	S20 24	178 07	S24 04
10	245 08.0	30.0	73 14.5	204 20		36 27		128 29		180 32	05
20	247 38.0	30.2	75 44.9	206 50		38 58		130 59		182 57	05
30	250 08.0	· 30.3	78 15.3	209 20	·	41 28	·	133 29	·	185 22	· 05
40	252 38.0	30.5	80 45.7	211 50		43 59		136 00		187 47	05
50	255 08.1	30.7	83 16.1	214 20		46 29		138 30		190 12	05
05 00	257 38.1	S 3 30.8	85 46.6	216 50	S19 58	49 00	N11 22	141 01	S20 24	192 36	S24 05
10	260 08.1	31.0	88 17.0	219 19		51 30		143 31		195 01	06
20	262 38.2	31.2	90 47.4	221 49		54 01		146 01		197 26	06
30	265 08.2	· 31.3	93 17.8	224 19	·	56 31	·	148 32	·	199 51	· 06
40	267 38.2	31.5	95 48.2	226 49		59 02		151 02		202 16	06
50	270 08.3	31.6	98 18.6	229 19		61 32		153 33		204 41	06
06 00	272 38.3	S 3 31.8	100 49.0	231 49	S19 59	64 03	N11 22	156 03	S20 24	207 06	S24 06
10	275 08.3	32.0	103 19.4	234 19		66 33		158 33		209 31	06
20	277 38.4	32.1	105 49.8	236 49		69 04		161 04		211 56	07
30	280 08.4	· 32.3	108 20.2	239 19	·	71 34	·	163 34	·	214 21	· 07
40	282 38.4	32.4	110 50.7	241 49		74 05		166 05		216 46	07
50	285 08.5	32.6	113 21.1	244 19		76 35		168 35		219 11	07
07 00	287 38.5	S 3 32.8	115 51.5	246 49	S19 59	79 06	N11 22	171 05	S20 24	221 36	S24 07
10	290 08.5	32.9	118 21.9	249 19		81 36		173 36		224 01	07
20	292 38.6	33.1	120 52.3	251 48		84 07		176 06		226 26	07
30	295 08.6	· 33.3	123 22.7	254 18	·	86 37	·	178 37	·	228 51	· 07
40	297 38.6	33.4	125 53.1	256 48		89 08		181 07		231 16	07
50	300 08.7	33.6	128 23.5	259 18		91 38		183 37		233 41	08
08 00	302 38.7	S 3 33.7	130 53.9	261 48	S20 00	94 09	N11 22	186 08	S20 24	236 06	S24 08
10	305 08.7	33.9	133 24.4	264 18		96 39		188 38		238 31	08
20	307 38.8	34.1	135 54.8	266 48		99 10		191 09		240 56	08
30	310 08.8	· 34.2	138 25.2	269 18	·	101 40	·	193 39	·	243 20	· 08
40	312 38.8	34.4	140 55.6	271 48		104 11		196 09		245 45	08
50	315 08.9	34.5	143 26.0	274 18		106 41		198 40		248 10	08
09 00	317 38.9	S 3 34.7	145 56.4	276 48	S20 01	109 12	N11 22	201 10	S20 24	250 35	S24 08
10	320 08.9	34.9	148 26.8	279 18		111 42		203 41		253 00	08
20	322 39.0	35.0	150 57.2	281 48		114 12		206 11		255 25	08
30	325 09.0	· 35.2	153 27.6	284 18	·	116 43	·	208 41	·	257 50	· 08
40	327 39.0	35.4	155 58.1	286 47		119 13		211 12		260 15	09
50	330 09.1	35.5	158 28.5	289 17		121 44		213 42		262 40	09
10 00	332 39.1	S 3 35.7	160 58.9	291 47	S20 02	124 14	N11 22	216 13	S20 24	265 05	S24 09
10	335 09.1	35.8	163 29.3	294 17		126 45		218 43		267 30	09
20	337 39.2	36.0	165 59.7	296 47		129 15		221 14		269 55	09
30	340 09.2	· 36.2	168 30.1	299 17	·	131 46	·	223 44	·	272 20	· 09
40	342 39.2	36.3	171 00.5	301 47		134 16		226 14		274 45	09
50	345 09.3	36.5	173 30.9	304 17		136 47		228 45		277 10	09
11 00	347 39.3	S 3 36.6	176 01.3	306 47	S20 03	139 17	N11 22	231 15	S20 24	279 35	S24 09
10	350 09.3	36.8	178 31.7	309 17		141 48		233 46		282 00	09
20	352 39.4	37.0	181 02.2	311 47		144 18		236 16		284 25	09
30	355 09.4	· 37.1	183 32.6	314 17	·	146 49	·	238 46	·	286 50	· 09
40	357 39.4	37.3	186 03.0	316 47		149 19		241 17		289 15	09
50	0 09.5	37.5	188 33.4	319 16		151 50		243 47		291 40	09
Rate	15 00.2	S0 01.0		14 59.6	S0 00.9	15 02.9	S0 00.1	15 02.4	0 00.0	14 29.7	S0 00.8*

Lat.	Moon-rise	Diff.
N		
°	h m	m
72	■	*
70	■	*
68	■	*
66	■	*
64	14 59	24
62	14 20	25
60	13 53	25
58	13 31	25
56	13 14	25
54	12 59	25
52	12 46	25
50	12 35	25
45	12 11	25
40	11 52	25
35	11 37	25
30	11 23	25
20	11 00	25
10	10 40	25
0	10 21	25
10	10 02	25
20	09 42	25
30	09 19	25
35	09 05	25
40	08 50	25
45	08 31	25
50	08 08	24
52	07 56	24
54	07 44	24
56	07 29	24
58	07 12	23
60	06 51	23
S		

Moon's P. in A.

Alt. °	Corr. + ′	Alt. °	Corr. + ′
0		57	
	54		29
10		58	
	53		28
15		59	
	52		27
19		60	
	51		26
22		62	
	50		25
24		63	
	49		24
27		64	
	48		23
29		65	
	47		22
31		66	
	46		21
33		67	
	45		20
35		69	
	44		19
37		70	
	43		18
38		71	
	42		17
40		72	
	41		16
41		73	
	40		15
43		74	
	39		14
45		75	
	38		13
46		76	
	37		12
47		77	
	36		11
49		78	
	35		10
50		79	
	34		
52			
	33		
53			
	32		
54			
	31		
55			
	30		
57			
	29		
58			

Sun SD 16′.0

Moon SD 15′

Age 6d

TABLE VII Air Almanac page 549 Greenwich A.M. 1973 October 2

A84 *POLARIS* (POLE STAR) TABLE, 1973

FOR DETERMINING THE LATITUDE FROM A SEXTANT ALTITUDE

L.H.A. ♈ ° ′	Q ′	L.H.A. ♈ ° ′	Q ′	L.H.A. ♈ ° ′	Q ′	L.H.A. ♈ ° ′	Q ′	L.H.A. ♈ ° ′	Q ′	L.H.A. ♈ ° ′	Q ′	L.H.A. ♈ ° ′	Q ′	L.H.A. ♈ ° ′	Q ′
359 49	−44	**81 57**	−32	**113 53**	− 6	**143 33**	+20	**183 54**	+46	**265 27**	+30	**296 59**	+ 4	**326 46**	−22
1 59	−45	**83 22**	−31	**115 00**	− 5	**144 46**	+21	**186 26**	+47	**266 50**	+29	**298 06**	+ 3	**328 00**	−23
4 17	−46	**84 46**	−30	**116 08**	− 4	**145 59**	+22	**189 15**	+48	**268 12**	+28	**299 13**	+ 2	**329 14**	−24
6 48	−47	**86 08**	−29	**117 15**	− 3	**147 14**	+23	**192 26**	+49	**269 32**	+27	**300 20**	+ 1	**330 30**	−25
9 34	−48	**87 29**	−28	**118 22**	− 2	**148 29**	+24	**196 16**	+50	**270 51**	+26	**301 27**	0	**331 47**	−26
12 43	−49	**88 48**	−27	**119 29**	− 1	**149 45**	+25	**201 24**	+51	**272 09**	+25	**302 35**	− 1	**333 04**	−27
16 29	−50	**90 07**	−26	**120 36**	0	**151 02**	+26	**221 47**	+50	**273 26**	+24	**303 42**	− 2	**334 23**	−28
21 33	−51	**91 24**	−25	**121 44**	+ 1	**152 20**	+27	**226 55**	+49	**274 42**	+23	**304 49**	− 3	**335 42**	−29
41 38	−50	**92 41**	−24	**122 51**	+ 2	**153 39**	+28	**230 45**	+48	**275 57**	+22	**305 56**	− 4	**337 03**	−30
46 42	−49	**93 57**	−23	**123 58**	+ 3	**154 59**	+29	**233 56**	+47	**277 12**	+21	**307 03**	− 5	**338 25**	−31
50 28	−48	**95 11**	−22	**125 05**	+ 4	**156 21**	+30	**236 45**	+46	**278 25**	+20	**308 11**	− 6	**339 49**	−32
53 37	−47	**96 25**	−21	**126 12**	+ 5	**157 44**	+31	**239 17**	+45	**279 38**	+19	**309 18**	− 7	**341 14**	−33
56 23	−46	**97 38**	−20	**127 20**	+ 6	**159 09**	+32	**241 38**	+44	**280 51**	+18	**310 26**	− 8	**342 41**	−34
58 54	−45	**98 51**	−19	**128 27**	+ 7	**160 35**	+33	**243 49**	+43	**282 02**	+17	**311 33**	− 9	**344 11**	−35
61 12	−44	**100 03**	−18	**129 35**	+ 8	**162 03**	+34	**245 53**	+42	**283 14**	+16	**312 41**	−10	**345 42**	−36
63 22	−43	**101 14**	−17	**130 43**	+ 9	**163 33**	+35	**247 51**	+41	**284 24**	+15	**313 50**	−11	**347 15**	−37
65 24	−42	**102 25**	−16	**131 51**	+10	**165 05**	+36	**249 43**	+40	**285 35**	+14	**314 58**	−12	**348 52**	−38
67 20	−41	**103 35**	−15	**132 59**	+11	**166 39**	+37	**251 30**	+39	**286 44**	+13	**316 07**	−13	**350 30**	−39
69 11	−40	**104 45**	−14	**134 08**	+12	**168 16**	+38	**253 14**	+38	**287 54**	+12	**317 16**	−14	**352 13**	−40
70 58	−39	**105 55**	−13	**135 17**	+13	**169 57**	+39	**254 55**	+37	**289 03**	+11	**318 26**	−15	**354 00**	−41
72 41	−38	**107 04**	−12	**136 27**	+14	**171 41**	+40	**256 32**	+36	**290 12**	+10	**319 36**	−16	**355 51**	−42
74 19	−37	**108 13**	−11	**137 36**	+15	**173 28**	+41	**258 06**	+35	**291 20**	+ 9	**320 46**	−17	**357 47**	−43
75 56	−36	**109 21**	−10	**138 47**	+16	**175 20**	+42	**259 38**	+34	**292 28**	+ 8	**321 57**	−18	**359 49**	−44
77 29	−35	**110 30**	− 9	**139 57**	+17	**177 18**	+43	**261 08**	+33	**293 36**	+ 7	**323 08**	−19	**1 59**	−45
79 00	−34	**111 38**	− 8	**141 09**	+18	**179 22**	+44	**262 36**	+32	**294 44**	+ 6	**324 20**	−20	**4 17**	−46
80 30	−33	**112 45**	− 7	**142 20**	+19	**181 33**	+45	**264 02**	+31	**295 51**	+ 5	**325 33**	−21	**6 48**	−47
81 57		**113 53**		**143 33**		**183 54**		**265 27**		**296 59**		**326 46**		**9 34**	

Q, which does *not* include refraction, is to be applied to the corrected sextant altitude of *Polaris*.
Polaris: Mag. 2·1, S.H.A. 328° 24′, Dec. N. 89° 08′·7

TABLE VIII Air Almanac page A84 Polaris (Pole Star) Table, 1973

ADDENDUM

GMT	☉ SUN GHA	SUN Dec.	ARIES GHA ♈	VENUS −3.7 GHA	VENUS Dec.	MARS −1.9 GHA	MARS Dec.	JUPITER −2.1 GHA	JUPITER Dec.	◐ MOON GHA	MOON Dec.
h m	° ′	° ′	° ′	° ′	° ′	° ′	° ′	° ′	° ′	° ′	° ′
12 00	2 39.5	S 3 37.6	191 03.8	321 46	S20 04	154 20	N11 22	246 18	S20 24	294 05	S24 09
10	5 09.5	37.8	193 34.2	324 16		156 51		248 48		296 30	09
20	7 39.6	37.9	196 04.6	326 46		159 21		251 18		298 55	09
30	10 09.6	38.1	198 35.0	329 16		161 52		253 49		301 19	09
40	12 39.6	38.3	201 05.4	331 46		164 22		256 19		303 44	09
50	15 09.7	38.4	203 35.9	334 16		166 53		258 50		306 09	09
13 00	17 39.7	S 3 38.6	206 06.3	336 46	S20 05	169 23	N11 22	261 20	S20 24	308 34	S24 09
10	20 09.7	38.7	208 36.7	339 16		171 54		263 50		310 59	09
20	22 39.8	38.9	211 07.1	341 46		174 24		266 21		313 24	09
30	25 09.8	39.1	213 37.5	344 16		176 55		268 51		315 49	09
40	27 39.8	39.2	216 07.9	346 46		179 25		271 22		318 14	09
50	30 09.9	39.4	218 38.3	349 16		181 56		273 52		320 39	09
14 00	32 39.9	S 3 39.6	221 08.7	351 46	S20 06	184 26	N11 22	276 22	S20 24	323 04	S24 09
10	35 09.9	39.7	223 39.1	354 15		186 57		278 53		325 29	09
20	37 40.0	39.9	226 09.6	356 45		189 27		281 23		327 54	09
30	40 10.0	40.0	228 40.0	359 15		191 58		283 54		330 19	09
40	42 40.0	40.2	231 10.4	1 45		194 28		286 24		332 44	09
50	45 10.1	40.4	233 40.8	4 15		196 59		288 54		335 09	09
15 00	47 40.1	S 3 40.5	236 11.2	6 45	S20 07	199 29	N11 22	291 25	S20 24	337 34	S24 09
10	50 10.1	40.7	238 41.6	9 15		202 00		293 55		339 59	09
20	52 40.2	40.8	241 12.0	11 45		204 30		296 26		342 24	09
30	55 10.2	41.0	243 42.4	14 15		207 01		298 56		344 49	09
40	57 40.2	41.2	246 12.8	16 45		209 31		301 26		347 14	09
50	60 10.3	41.3	248 43.2	19 15		212 02		303 57		349 39	09
16 00	62 40.3	S 3 41.5	251 13.7	21 45	S20 08	214 32	N11 22	306 27	S20 24	352 04	S24 09
10	65 10.3	41.6	253 44.1	24 15		217 03		308 58		354 29	09
20	67 40.4	41.8	256 14.5	26 45		219 33		311 28		356 54	08
30	70 10.4	42.0	258 44.9	29 14		222 04		313 58		359 19	08
40	72 40.4	42.1	261 15.3	31 44		224 34		316 29		1 44	08
50	75 10.5	42.3	263 45.7	34 14		227 05		318 59		4 09	08
17 00	77 40.5	S 3 42.5	266 16.1	36 44	S20 08	229 35	N11 22	321 30	S20 24	6 34	S24 08
10	80 10.5	42.6	268 46.5	39 14		232 05		324 00		8 59	08
20	82 40.6	42.8	271 16.9	41 44		234 36		326 31		11 24	08
30	85 10.6	42.9	273 47.4	44 14		237 06		329 01		13 49	08
40	87 40.6	43.1	276 17.8	46 44		239 37		331 31		16 14	08
50	90 10.6	43.3	278 48.2	49 14		242 07		334 02		18 39	08
18 00	92 40.7	S 3 43.4	281 18.6	51 44	S20 09	244 38	N11 22	336 32	S20 24	21 03	S24 08
10	95 10.7	43.6	283 49.0	54 14		247 08		339 03		23 28	08
20	97 40.7	43.7	286 19.4	56 44		249 39		341 33		25 53	07
30	100 10.8	43.9	288 49.8	59 14		252 09		344 03		28 18	07
40	102 40.8	44.1	291 20.2	61 43		254 40		346 34		30 43	07
50	105 10.8	44.2	293 50.6	64 13		257 10		349 04		33 08	07
19 00	107 40.9	S 3 44.4	296 21.1	66 43	S20 10	259 41	N11 22	351 35	S20 24	35 33	S24 07
10	110 10.9	44.6	298 51.5	69 13		262 11		354 05		37 58	07
20	112 40.9	44.7	301 21.9	71 43		264 42		356 35		40 23	07
30	115 11.0	44.9	303 52.3	74 13		267 12		359 06		42 48	07
40	117 41.0	45.0	306 22.7	76 43		269 43		1 36		45 13	07
50	120 11.0	45.2	308 53.1	79 13		272 13		4 07		47 38	06
20 00	122 41.1	S 3 45.4	311 23.5	81 43	S20 11	274 44	N11 21	6 37	S20 24	50 03	S24 06
10	125 11.1	45.5	313 53.9	84 13		277 14		9 07		52 28	06
20	127 41.1	45.7	316 24.3	86 43		279 45		11 38		54 53	06
30	130 11.2	45.8	318 54.7	89 13		282 15		14 08		57 18	06
40	132 41.2	46.0	321 25.2	91 43		284 46		16 39		59 43	06
50	135 11.2	46.2	323 55.6	94 12		287 16		19 09		62 08	06
21 00	137 41.3	S 3 46.3	326 26.0	96 42	S20 12	289 47	N11 21	21 39	S20 24	64 33	S24 05
10	140 11.3	46.5	328 56.4	99 12		292 17		24 10		66 58	05
20	142 41.3	46.6	331 26.8	101 42		294 48		26 40		69 23	05
30	145 11.4	46.8	333 57.2	104 12		297 18		29 11		71 48	05
40	147 41.4	47.0	336 27.6	106 42		299 49		31 41		74 13	05
50	150 11.4	47.1	338 58.0	109 12		302 19		34 11		76 38	05
22 00	152 41.5	S 3 47.3	341 28.4	111 42	S20 13	304 50	N11 21	36 42	S20 24	79 03	S24 04
10	155 11.5	47.5	343 58.9	114 12		307 20		39 12		81 28	04
20	157 41.5	47.6	346 29.3	116 42		309 51		41 43		83 53	04
30	160 11.6	47.8	348 59.7	119 12		312 21		44 13		86 18	04
40	162 41.6	47.9	351 30.1	121 42		314 52		46 43		88 43	04
50	165 11.6	48.1	354 00.5	124 12		317 22		49 14		91 08	04
23 00	167 41.7	S 3 48.3	356 30.9	126 42	S20 14	319 53	N11 21	51 44	S20 24	93 33	S24 03
10	170 11.7	48.4	359 01.3	129 11		322 23		54 15		95 58	03
20	172 41.7	48.6	1 31.7	131 41		324 54		56 45		98 23	03
30	175 11.8	48.7	4 02.1	134 11		327 24		59 15		100 48	03
40	177 41.8	48.9	6 32.5	136 41		329 55		61 46		103 13	03
50	180 11.8	49.1	9 03.0	139 11		332 25		64 16		105 38	02
Rate	15 00.2	S0 01.0		14 59.6	S0 00.9	15 02.9	S0 00.1	15 02.4	0 00.0	14 29.9	N0 00.6

Lat.	Moon-set	Diff.
N		
°	h m	m
72	■	*
70	■	*
68	■	*
66	■	*
64	18 07	28
62	18 46	27
60	19 14	27
58	19 35	27
56	19 52	27
54	20 07	26
52	20 20	26
50	20 31	26
45	20 55	26
40	21 14	26
35	21 29	26
30	21 43	26
20	22 06	25
10	22 26	25
0	22 45	25
10	23 04	25
20	23 24	25
30	23 47	24
35	24 00	24
40	24 16	24
45	24 34	23
50	00 07	27
52	00 18	27
54	00 31	28
56	00 45	28
58	01 02	28
60	01 23	29
S		

Moon's P. in A.

Alt.	Corr.	Alt.	Corr.
°	+ ′	°	+ ′
0		57	
	54		29
10		58	
	53		28
15		59	
	52		27
18		60	
	51		26
21		62	
	50		25
24		63	
	49		24
26		64	
	48		23
29		65	
	47		22
31		66	
	46		21
33		67	
	45		20
35		68	
	44		19
36		70	
	43		18
38		71	
	42		17
40		72	
	41		16
41		73	
	40		15
43		74	
	39		14
44		75	
	38		13
46		76	
	37		12
47		77	
	36		11
49		78	
	35		10
50		79	
	34		
51			
	33		
53			
	32		
54			
	31		
55			
	30		
57			
	29		
58			

Sun SD 16′.0
Moon SD 15′
Age 6d

Addendum 1 Air Almanac page 550 Greenwich P.M. 1973 October 2

GMT	☉ SUN GHA	SUN Dec.	ARIES GHA ♈	VENUS −3.7 GHA	VENUS Dec.	MARS −1.9 GHA	MARS Dec.	JUPITER −2.1 GHA	JUPITER Dec.	◐ MOON GHA	MOON Dec.
h m	° ′	° ′	° ′	° ′	° ′	° ′	° ′	° ′	° ′	° ′	° ′
00 00	182 41.9	S 3 49.2	11 33.4	141 41	S20 15	334 56	N11 21	66 47	S20 24	108 03	S24 02
10	185 11.9	49.4	14 03.8	144 11		337 26		69 17		110 28	02
20	187 41.9	49.6	16 34.2	146 41		339 57		71 48		112 53	02
30	190 12.0	· 49.7	19 04.6	149 11	· ·	342 27	· ·	74 18	· ·	115 18	· 01
40	192 42.0	49.9	21 35.0	151 41		344 58		76 48		117 43	01
50	195 12.0	50.0	24 05.4	154 11		347 28		79 19		120 08	01
01 00	197 42.1	S 3 50.2	26 35.8	156 41	S20 16	349 59	N11 21	81 49	S20 24	122 33	S24 01
10	200 12.1	50.4	29 06.2	159 11		352 29		84 20		124 58	01
20	202 42.1	50.5	31 36.7	161 41		355 00		86 50		127 23	00
30	205 12.2	· 50.7	34 07.1	164 10	· ·	357 30	· ·	89 20	· ·	129 48	· 00
40	207 42.2	50.8	36 37.5	166 40		0 01		91 51		132 13	00
50	210 12.2	51.0	39 07.9	169 10		2 31		94 21		134 38	24 00
02 00	212 42.3	S 3 51.2	41 38.3	171 40	S20 16	5 02	N11 21	96 52	S20 24	137 03	S23 59
10	215 12.3	51.3	44 08.7	174 10		7 32		99 22		139 28	59
20	217 42.3	51.5	46 39.1	176 40		10 02		101 52		141 53	59
30	220 12.3	· 51.6	49 09.5	179 10	· ·	12 33	· ·	104 23	· ·	144 18	· 59
40	222 42.4	51.8	51 39.9	181 40		15 03		106 53		146 43	58
50	225 12.4	52.0	54 10.4	184 10		17 34		109 24		149 08	58
03 00	227 42.4	S 3 52.1	56 40.8	186 40	S20 17	20 04	N11 21	111 54	S20 24	151 33	S23 58
10	230 12.5	52.3	59 11.2	189 10		22 35		114 24		153 58	58
20	232 42.5	52.5	61 41.6	191 40		25 05		116 55		156 23	57
30	235 12.5	· 52.6	64 12.0	194 10	· ·	27 36	· ·	119 25	· ·	158 48	· 57
40	237 42.6	52.8	66 42.4	196 39		30 06		121 56		161 13	57
50	240 12.6	52.9	69 12.8	199 09		32 37		124 26		163 38	56
04 00	242 42.6	S 3 53.1	71 43.2	201 39	S20 18	35 07	N11 21	126 56	S20 24	166 03	S23 56
10	245 12.7	53.3	74 13.6	204 09		37 38		129 27		168 28	56
20	247 42.7	53.4	76 44.0	206 39		40 08		131 57		170 53	56
30	250 12.7	· 53.6	79 14.5	209 09	· ·	42 39	· ·	134 28	· ·	173 18	· 55
40	252 42.8	53.7	81 44.9	211 39		45 09		136 58		175 43	55
50	255 12.8	53.9	84 15.3	214 09		47 40		139 28		178 08	55
05 00	257 42.8	S 3 54.1	86 45.7	216 39	S20 19	50 10	N11 21	141 59	S20 24	180 33	S23 54
10	260 12.9	54.2	89 16.1	219 09		52 41		144 29		182 58	54
20	262 42.9	54.4	91 46.5	221 39		55 11		147 00		185 23	54
30	265 12.9	· 54.5	94 16.9	224 09	· ·	57 42	· ·	149 30	· ·	187 48	· 53
40	267 43.0	54.7	96 47.3	226 39		60 12		152 00		190 13	53
50	270 13.0	54.9	99 17.7	229 08		62 43		154 31		192 38	53
06 00	272 43.0	S 3 55.0	101 48.2	231 38	S20 20	65 13	N11 21	157 01	S20 24	195 03	S23 53
10	275 13.1	55.2	104 18.6	234 08		67 44		159 32		197 28	52
20	277 43.1	55.4	106 49.0	236 38		70 14		162 02		199 53	52
30	280 13.1	· 55.5	109 19.4	239 08	· ·	72 45	· ·	164 32	· ·	202 18	· 52
40	282 43.2	55.7	111 49.8	241 38		75 15		167 03		204 43	51
50	285 13.2	55.8	114 20.2	244 08		77 46		169 33		207 08	51
07 00	287 43.2	S 3 56.0	116 50.6	246 38	S20 21	80 16	N11 21	172 04	S20 24	209 33	S23 51
10	290 13.3	56.2	119 21.0	249 08		82 47		174 34		211 58	50
20	292 43.3	56.3	121 51.4	251 38		85 17		177 04		214 23	50
30	295 13.3	· 56.5	124 21.9	254 08	· ·	87 48	· ·	179 35	· ·	216 48	· 49
40	297 43.4	56.6	126 52.3	256 38		90 18		182 05		219 13	49
50	300 13.4	56.8	129 22.7	259 08		92 49		184 36		221 38	49
08 00	302 43.4	S 3 57.0	131 53.1	261 38	S20 22	95 19	N11 21	187 06	S20 24	224 03	S23 48
10	305 13.5	57.1	134 23.5	264 07		97 50		189 37		226 28	48
20	307 43.5	57.3	136 53.9	266 37		100 20		192 07		228 53	48
30	310 13.5	· 57.4	139 24.3	269 07	· ·	102 51	· ·	194 37	· ·	231 18	· 47
40	312 43.6	57.6	141 54.7	271 37		105 21		197 08		233 43	47
50	315 13.6	57.8	144 25.1	274 07		107 52		199 38		236 09	47
09 00	317 43.6	S 3 57.9	146 55.5	276 37	S20 23	110 22	N11 21	202 09	S20 24	238 34	S23 46
10	320 13.6	58.1	149 26.0	279 07		112 53		204 39		240 59	46
20	322 43.7	58.3	151 56.4	281 37		115 23		207 09		243 24	45
30	325 13.7	· 58.4	154 26.8	284 07	· ·	117 54	· ·	209 40	· ·	245 49	· 45
40	327 43.7	58.6	156 57.2	286 37		120 24		212 10		248 14	45
50	330 13.8	58.7	159 27.6	289 07		122 55		214 41		250 39	44
10 00	332 43.8	S 3 58.9	161 58.0	291 37	S20 23	125 25	N11 21	217 11	S20 24	253 04	S23 44
10	335 13.8	59.1	164 28.4	294 07		127 56		219 41		255 29	43
20	337 43.9	59.2	166 58.8	296 36		130 26		222 12		257 54	43
30	340 13.9	· 59.4	169 29.2	299 06	· ·	132 57	· ·	224 42	· ·	260 19	· 43
40	342 43.9	59.5	171 59.7	301 36		135 27		227 13		262 44	42
50	345 14.0	59.7	174 30.1	304 06		137 58		229 43		265 09	42
11 00	347 44.0	S 3 59.9	177 00.5	306 36	S20 24	140 28	N11 20	232 13	S20 24	267 34	S23 41
10	350 14.0	4 00.0	179 30.9	309 06		142 59		234 44		269 59	41
20	352 44.1	00.2	182 01.3	311 36		145 29		237 14		272 24	41
30	355 14.1	· 00.3	184 31.7	314 06	· ·	148 00	· ·	239 45	· ·	274 49	· 40
40	357 44.1	00.5	187 02.1	316 36		150 30		242 15		277 14	40
50	0 14.2	00.7	189 32.5	319 06		153 01		244 45		279 39	39
Rate	15 00.2	S0 01.0		14 59.6	S0 00.9	15 03.0	S0 00.1	15 02.4	0 00.0	14 30.1	N0 01.9*

Lat.	Moonrise	Diff.
°	h m	m
N		
72	■	*
70	■	*
68	■	*
66	16 32	*
64	15 33	12
62	14 59	16
60	14 34	17
58	14 14	19
56	13 58	19
54	13 44	20
52	13 31	21
50	13 21	21
45	12 58	22
40	12 40	22
35	12 24	23
30	12 11	23
20	11 48	24
10	11 29	24
0	11 10	25
10	10 52	25
20	10 32	26
30	10 10	26
35	09 56	27
40	09 41	27
45	09 23	27
50	09 00	28
52	08 49	28
54	08 36	29
56	08 22	29
58	08 05	30
60	07 45	30
S		

Moon's P. in A.

Alt. °	Corr. + ′	Alt. °	Corr. + ′
0	54	57	29
9	53	58	28
14	52	59	27
18	51	60	26
21	50	61	25
24	49	63	24
26	48	64	23
28	47	65	22
31	46	66	21
33	45	67	20
34	44	68	19
36	43	70	18
38	42	71	17
40	41	72	16
41	40	73	15
43	39	74	14
44	38	75	13
46	37	76	12
47	36	77	11
49	35	78	10
50	34	79	
51	33		
53	32		
54	31		
55	30		
57	29		
58			

Sun SD 16′.0
Moon SD 15′
Age 7d

Addendum 2 Air Almanac page 551 Greenwich A.M. 1973 October 3

N. Lat. {LHA greater than 180°........ Zn=Z; LHA less than 180°............ Zn=360−Z

DECLINATION (15°-29°) CONTRARY NAME TO LATITUDE

LHA	15° Hc	d	Z	16° Hc	d	Z	17° Hc	d	Z	18° Hc	d	Z	19° Hc	d	Z
69	08 35	36	114	07 59	37	115	07 22	37	116	06 45	37	117	06 08	37	117
68	09 21	37	115	08 44	37	116	08 07	37	116	07 30	37	117	06 53	38	118
67	10 07	38	115	09 29	37	116	08 52	37	117	08 15	38	118	07 37	38	119
66	10 52	38	116	10 14	37	117	09 37	38	118	08 59	38	118	08 21	38	119
65	11 37	−38	117	10 59	−38	117	10 21	−38	118	09 43	−38	119	09 05	−39	120
64	12 22	38	117	11 44	39	118	11 05	38	119	10 27	39	120	09 48	38	120
63	13 07	39	118	12 28	39	119	11 49	38	120	11 11	39	120	10 32	39	121
62	13 51	39	119	13 12	39	119	12 33	39	120	11 54	39	121	11 15	40	122
61	14 35	39	119	13 56	40	120	13 16	39	121	12 37	40	122	11 57	39	122
60	15 19	−40	120	14 39	−39	121	14 00	−40	121	13 20	−40	122	12 40	−40	123
59	16 02	40	121	15 22	40	121	14 42	40	122	14 02	40	123	13 22	41	124
58	16 45	40	121	16 05	40	122	15 25	41	123	14 44	40	124	14 04	41	124
57	17 28	40	122	16 48	41	123	16 07	41	123	15 26	41	124	14 45	41	125
56	18 11	41	123	17 30	41	123	16 49	41	124	16 08	42	125	15 26	42	126
55	18 53	−41	123	18 12	−42	124	17 30	−41	125	16 49	−42	126	16 07	−42	126
54	19 35	42	124	18 53	42	125	18 11	42	126	17 29	42	126	16 47	42	127
53	20 17	42	125	19 35	43	125	18 52	42	126	18 10	43	127	17 27	43	128
52	20 58	43	125	20 15	42	126	19 33	43	127	18 50	43	128	18 07	43	128
51	21 39	43	126	20 56	43	127	20 13	43	128	19 30	44	128	18 46	44	129
50	22 19	−43	127	21 36	−44	128	20 52	−43	128	20 09	−44	129	19 25	−44	130
49	22 59	43	128	22 16	44	128	21 32	44	129	20 48	45	130	20 03	44	131
48	23 39	44	128	22 55	45	129	22 10	44	130	21 26	45	131	20 41	44	131
47	24 18	44	129	23 34	45	130	22 49	45	131	22 04	45	131	21 19	45	132
46	24 57	45	130	24 12	45	131	23 27	45	131	22 42	46	132	21 56	46	133
45	25 35	−45	131	24 50	−46	132	24 04	−45	132	23 19	−46	133	22 33	−46	134
44	26 13	46	132	25 27	46	132	24 41	46	133	23 55	46	134	23 09	47	134
43	26 50	46	132	26 04	46	133	25 18	47	134	24 31	46	135	23 45	47	135
42	27 27	46	133	26 41	47	134	25 54	47	135	25 07	47	135	24 20	48	136
41	28 04	47	134	27 17	47	135	26 30	48	136	25 42	48	136	24 54	47	137
40	28 40	−48	135	27 52	−47	136	27 05	−48	136	26 17	−48	137	25 29	−49	138
39	29 15	48	136	28 27	48	137	27 39	48	137	26 51	49	138	26 02	48	139
38	29 50	49	137	29 01	48	137	28 13	49	138	27 24	49	139	26 35	49	139
37	30 24	49	138	29 35	49	138	28 46	49	139	27 57	49	140	27 08	50	140
36	30 58	50	139	30 08	49	139	29 19	50	140	28 29	49	141	27 40	50	141
35	31 31	−50	140	30 41	−50	140	29 51	−50	141	29 01	−50	141	28 11	−50	142
34	32 03	50	140	31 13	50	141	30 23	51	142	29 32	50	142	28 42	51	143
33	32 35	51	141	31 44	51	142	30 53	50	143	30 03	51	143	29 12	52	144
32	33 06	51	142	32 15	51	143	31 24	52	144	30 32	51	144	29 41	52	145
31	33 36	51	143	32 45	52	144	31 53	52	145	31 01	51	145	30 10	52	146
30	34 06	−52	144	33 14	−52	145	32 22	−52	146	31 30	−52	146	30 38	−53	147
29	34 35	52	145	33 43	53	146	32 50	52	147	31 58	53	147	31 05	53	148
28	35 03	53	146	34 10	52	147	33 18	53	148	32 25	54	148	31 31	53	149
27	35 31	54	147	34 37	53	148	33 44	53	149	32 51	54	149	31 57	53	150
26	35 57	53	149	35 04	54	149	34 10	54	150	33 16	54	150	32 22	54	151
25	36 23	−54	150	35 29	−54	150	34 35	−54	151	33 41	−54	151	32 47	−55	152
24	36 48	54	151	35 54	54	151	35 00	55	152	34 05	55	152	33 10	55	153
23	37 13	55	152	36 18	55	152	35 23	55	153	34 28	55	153	33 33	55	154
22	37 36	55	153	36 41	55	153	35 46	56	154	34 50	55	154	33 55	56	155
21	37 59	56	154	37 03	56	154	36 07	55	155	35 12	56	155	34 16	56	156
20	38 20	−56	155	37 24	−56	156	36 28	−56	156	35 32	−56	156	34 36	−56	157
19	38 41	56	156	37 45	57	157	36 48	56	157	35 52	57	158	34 55	56	158
18	39 01	57	157	38 04	57	158	37 07	56	158	36 11	57	159	35 14	57	159
17	39 20	57	159	38 23	57	159	37 26	57	159	36 29	58	160	35 31	57	160
16	39 38	58	160	38 40	57	160	37 43	58	161	36 45	57	161	35 48	58	161
15	39 54	−57	161	38 57	−58	161	37 59	−58	162	37 01	−57	162	36 04	−58	162
14	40 10	58	162	39 12	58	163	38 14	58	163	37 16	58	163	36 18	58	164
13	40 25	58	163	39 27	58	164	38 29	58	164	37 31	59	164	36 32	58	165
12	40 39	58	165	39 41	59	165	38 42	58	165	37 44	59	166	36 45	58	166
11	40 52	59	166	39 53	59	166	38 54	58	166	37 56	59	167	36 57	59	167
10	41 04	−59	167	40 05	−59	167	39 06	−59	168	38 07	−59	168	37 08	−59	168
9	41 14	59	168	40 15	59	169	39 16	59	169	38 17	59	169	37 18	60	169
8	41 24	60	170	40 24	59	170	39 25	59	170	38 26	59	170	37 27	60	171
7	41 32	59	171	40 33	60	171	39 33	59	171	38 34	60	172	37 34	59	172
6	41 40	60	172	40 40	60	172	39 40	59	173	38 41	60	173	37 41	59	173
5	41 46	−60	174	40 46	−60	174	39 46	−59	174	38 47	−60	174	37 47	−60	174
4	41 51	60	175	40 51	60	175	39 51	60	175	38 51	59	175	37 52	60	175
3	41 55	60	176	40 55	60	176	39 55	60	176	38 55	60	176	37 55	60	176
2	41 58	60	177	40 58	60	178	39 58	60	178	38 58	60	178	37 58	60	178
1	41 59	60	179	40 59	59	179	40 00	60	179	39 00	60	179	38 00	60	179
0	42 00	−60	180	41 00	−60	180	40 00	−60	180	39 00	−60	180	38 00	−60	180

LHA	20° Hc	d	Z	21° Hc	d	Z	22° Hc	d	Z	23° Hc	d	Z	24° Hc	d	Z
69	05 31	37	118	04 54	37	119	04 17	37	120	03 40	38	121	03 02	37	121
68	06 15	37	119	05 38	38	120	05 00	37	120	04 23	38	121	03 45	37	122
67	06 59	37	119	06 22	38	120	05 44	38	121	05 06	38	122	04 28	38	122
66	07 43	38	120	07 05	38	121	06 27	38	122	05 49	39	122	05 10	38	123
65	08 26	−38	121	07 48	−38	121	07 10	−39	122	06 31	−39	123	05 52	−38	124
64	09 10	39	121	08 31	39	122	07 52	39	123	07 13	39	124	06 34	39	124
63	09 53	40	122	09 13	39	123	08 34	39	123	07 55	40	124	07 15	39	125
62	10 35	39	122	09 56	40	123	09 16	40	124	08 36	39	125	07 57	40	126
61	11 18	40	123	10 38	40	124	09 58	40	125	09 18	41	125	08 37	40	126
60	12 00	−41	124	11 19	−40	125	10 39	−41	125	09 58	−40	126	09 18	−41	127
59	12 41	40	124	12 01	41	125	11 20	41	126	10 39	41	127	09 58	41	127
58	13 23	41	125	12 42	41	126	12 01	42	127	11 19	41	127	10 38	42	128
57	14 04	42	126	13 22	41	126	12 41	42	127	11 59	42	128	11 17	41	129
56	14 44	41	126	14 03	42	127	13 21	42	128	12 39	42	129	11 57	43	129
55	15 25	−42	127	14 43	−43	128	14 00	−42	129	13 18	−43	129	12 35	−42	130
54	16 05	43	128	15 22	43	128	14 39	42	129	13 57	43	130	13 14	43	131
53	16 44	43	128	16 01	43	129	15 18	43	130	14 35	43	131	13 52	44	131
52	17 24	44	129	16 40	43	130	15 57	44	131	15 13	44	131	14 29	44	132
51	18 02	43	130	17 19	44	131	16 35	44	131	15 51	44	132	15 07	45	133
50	18 41	−44	131	17 57	−45	131	17 12	−44	132	16 28	−45	133	15 43	−44	133
49	19 19	45	131	18 34	44	132	17 50	45	133	17 05	45	133	16 20	45	134
48	19 57	45	132	19 12	46	133	18 26	45	133	17 41	45	134	16 56	46	135
47	20 34	46	133	19 48	45	134	19 03	46	134	18 17	46	135	17 31	46	136
46	21 10	45	134	20 25	46	134	19 39	47	135	18 52	46	136	18 06	46	136
45	21 47	−47	134	21 00	−46	135	20 14	−47	136	19 27	−46	136	18 41	−47	137
44	22 22	46	135	21 36	47	136	20 49	47	136	20 02	47	137	19 15	48	138
43	22 58	47	136	22 11	48	137	21 23	47	137	20 36	48	138	19 48	47	139
42	23 32	47	137	22 45	48	137	21 57	48	138	21 09	48	139	20 21	48	139
41	24 07	48	138	23 19	48	138	22 31	49	139	21 42	48	140	20 54	49	140
40	24 40	−48	138	23 52	−49	139	23 03	−48	140	22 15	−49	140	21 26	−49	141
39	25 14	49	139	24 25	49	140	23 36	49	141	22 47	50	141	21 57	49	142
38	25 46	49	140	24 57	49	141	24 08	50	141	23 18	50	142	22 28	49	143
37	26 18	49	141	25 29	50	142	24 39	50	142	23 49	50	143	22 59	51	143
36	26 50	50	142	26 00	51	142	25 09	50	143	24 19	51	144	23 28	50	144
35	27 21	−51	143	26 30	−51	143	25 39	−50	144	24 49	−51	144	23 58	−51	145
34	27 51	51	144	27 00	51	144	26 09	52	145	25 17	51	145	24 26	51	146
33	28 20	51	144	27 29	52	145	26 37	51	146	25 46	52	146	24 54	52	147
32	28 49	52	145	27 57	51	146	27 06	52	147	26 14	53	147	25 21	52	148
31	29 18	53	146	28 25	52	147	27 33	52	147	26 41	53	148	25 48	53	149
30	29 45	−52	147	28 53	−53	148	28 00	−53	148	27 07	−53	149	26 14	−53	149
29	30 12	53	148	29 19	53	149	28 26	53	149	27 33	54	150	26 39	53	150
28	30 38	53	149	29 45	54	150	28 51	53	150	27 58	54	151	27 04	54	151
27	31 04	54	150	30 10	54	151	29 16	54	151	28 22	54	152	27 28	55	152
26	31 28	54	151	30 34	54	152	29 40	55	152	28 45	54	153	27 51	55	153
25	31 52	−54	152	30 58	−55	153	30 03	−55	153	29 08	−55	154	28 13	−55	154
24	32 15	55	153	31 20	55	154	30 25	55	154	29 30	55	155	28 35	55	155
23	32 38	56	154	31 42	55	155	30 47	56	155	29 51	55	156	28 56	56	156
22	32 59	55	155	32 04	56	156	31 08	56	156	30 12	56	157	29 16	56	157
21	33 20	56	156	32 24	56	157	31 28	56	157	30 32	57	158	29 35	56	158
20	33 40	−57	157	32 43	−56	158	31 47	−57	158	30 50	−56	159	29 54	−57	159
19	33 59	57	158	33 02	57	159	32 05	57	159	31 08	56	160	30 12	57	160
18	34 17	57	159	33 20	57	160	32 23	57	160	31 26	58	161	30 28	57	161
17	34 34	57	161	33 37	58	161	32 39	57	161	31 42	57	162	30 45	58	162
16	34 50	57	162	33 53	58	162	32 55	58	162	31 57	57	163	31 00	58	163
15	35 06	−58	163	34 08	−58	163	33 10	−58	163	32 12	−59	164	31 14	−58	164
14	35 20	58	164	34 22	58	164	33 24	58	164	32 26	58	165	31 28	59	165
13	35 34	58	165	34 36	59	165	33 37	58	166	32 39	59	166	31 40	58	166
12	35 47	59	166	34 48	59	166	33 49	58	167	32 51	59	167	31 52	59	167
11	35 58	59	167	34 59	59	167	34 00	58	168	33 02	59	168	32 03	59	168
10	36 09	−59	168	35 10	−59	169	34 11	−59	169	33 12	−59	169	32 13	−60	169
9	36 18	59	170	35 19	59	170	34 20	59	170	33 21	59	170	32 22	60	170
8	36 27	59	171	35 28	60	171	34 28	59	171	33 29	59	171	32 30	60	171
7	36 35	60	172	35 35	59	172	34 36	60	172	33 36	59	172	32 37	60	172
6	36 42	60	173	35 42	60	173	34 42	59	173	33 43	60	173	32 43	60	174
5	36 47	−60	174	35 47	−59	174	34 48	−60	174	33 48	−60	175	32 48	−60	175
4	36 52	60	175	35 52	60	175	34 52	60	176	33 52	60	176	32 52	59	176
3	36 55	60	177	35 55	59	177	34 56	60	177	33 56	60	177	32 56	60	177
2	36 58	60	178	35 58	60	178	34 58	60	178	33 58	60	178	32 58	60	178
1	37 00	60	179	36 00	60	179	35 00	60	179	34 00	60	179	33 00	60	179
0	37 00	−60	180	36 00	−60	180	35 00	−60	180	34 00	−60	180	33 00	−60	180

LHA	25° Hc	d	Z	26° Hc	d	Z	27° Hc	d	Z	28° Hc	d	Z	29° Hc	d	Z	LHA
69	02 25	37	122	01 48	37	123	01 11	38	124	00 33	37	124	−0 04	37	125	291
68	03 08	38	123	02 30	38	123	01 52	37	124	01 15	38	125	00 37	38	126	292
67	03 50	38	123	03 12	38	124	02 34	38	125	01 56	38	126	01 18	39	126	293
66	04 32	39	124	03 53	38	125	03 15	39	125	02 36	38	126	01 58	39	127	294
65	05 14	−39	124	04 35	−39	125	03 56	−39	126	03 17	−39	127	02 38	−39	127	295
64	05 55	39	125	05 16	40	126	04 36	39	127	03 57	39	127	03 18	40	128	296
63	06 36	40	126	05 56	39	126	05 17	40	127	04 37	40	128	03 57	40	129	297
62	07 17	40	126	06 37	40	127	05 57	40	128	05 17	41	129	04 36	40	129	298
61	07 57	40	127	07 17	41	128	06 36	40	128	05 56	41	129	05 15	41	130	299
60	08 37	−41	128	07 56	−40	128	07 16	−41	129	06 35	−41	130	05 54	−41	130	300
59	09 17	41	128	08 36	41	129	07 55	42	130	07 13	41	130	06 32	42	131	301
58	09 56	41	129	09 15	42	129	08 33	41	130	07 52	42	131	07 10	42	132	302
57	10 36	42	129	09 54	42	130	09 12	43	131	08 29	42	132	07 47	42	132	303
56	11 14	42	130	10 32	43	131	09 49	42	131	09 07	43	132	08 24	42	133	304
55	11 53	−43	131	11 10	−43	131	10 27	−43	132	09 44	−43	133	09 01	−43	134	305
54	12 31	44	131	11 47	43	132	11 04	43	133	10 21	44	133	09 37	43	134	306
53	13 08	43	132	12 25	44	133	11 41	44	133	10 57	44	134	10 13	44	135	307
52	13 45	44	133	13 01	44	133	12 17	44	134	11 33	44	135	10 49	45	135	308
51	14 22	44	133	13 38	45	134	12 53	44	135	12 09	45	135	11 24	45	136	309
50	14 59	−45	134	14 14	−45	135	13 29	−45	135	12 44	−46	136	11 58	−45	137	310
49	15 35	46	135	14 49	45	135	14 04	46	136	13 18	45	137	12 33	46	138	311
48	16 10	46	136	15 24	46	136	14 38	45	137	13 53	47	138	13 06	46	138	312
47	16 45	46	136	15 59	46	137	15 13	47	138	14 26	46	138	13 40	47	139	313
46	17 20	47	137	16 33	47	138	15 46	46	138	15 00	47	139	14 13	47	140	314
45	17 54	−47	138	17 07	−47	138	16 20	−48	139	15 32	−47	140	14 45	−47	140	315
44	18 27	47	138	17 40	47	139	16 53	48	140	16 05	48	140	15 17	48	141	316
43	19 01	48	139	18 13	48	140	17 25	48	140	16 37	48	141	15 49	49	142	317
42	19 33	48	140	18 45	48	141	17 57	49	141	17 08	49	142	16 19	48	142	318
41	20 05	48	141	19 17	49	141	18 28	49	142	17 39	49	143	16 50	49	143	319
40	20 37	−49	142	19 48	−49	142	18 59	−50	143	18 09	−49	143	17 20	−50	144	320
39	21 08	50	142	20 18	49	143	19 29	50	144	18 39	50	144	17 49	50	145	321
38	21 39	50	143	20 49	51	144	19 58	50	144	19 08	50	145	18 18	50	145	322
37	22 08	50	144	21 18	50	145	20 28	51	145	19 37	51	146	18 46	51	146	323
36	22 38	51	145	21 47	51	145	20 56	51	146	20 05	51	147	19 14	51	147	324
35	23 07	−52	146	22 15	−51	146	21 24	−51	147	20 33	−52	147	19 41	−52	148	325
34	23 35	52	146	22 43	52	147	21 51	52	148	20 59	51	148	20 08	52	149	326
33	24 02	52	147	23 10	52	148	22 18	52	148	21 26	53	149	20 33	52	149	327
32	24 29	52	148	23 37	53	149	22 44	53	149	21 51	52	150	20 59	53	150	328
31	24 55	53	149	24 02	52	150	23 10	53	150	22 17	54	151	21 23	53	151	329
30	25 21	−53	150	24 28	−54	150	23 34	−53	151	22 41	−54	151	21 47	−53	152	330
29	25 46	54	151	24 52	54	151	23 58	53	152	23 05	54	152	22 11	54	153	331
28	26 10	54	152	25 16	54	152	24 22	54	153	23 28	54	153	22 34	55	154	332
27	26 33	54	153	25 39	54	153	24 45	55	154	23 50	54	154	22 56	55	155	333
26	26 56	55	154	26 01	54	154	25 07	55	154	24 12	55	155	23 17	55	155	334
25	27 18	−55	155	26 23	−55	155	25 28	−55	155	24 33	−55	156	23 38	−56	156	335
24	27 40	56	155	26 44	55	156	25 49	56	156	24 53	56	157	23 57	55	157	336
23	28 00	56	156	27 04	55	157	26 09	56	157	25 13	56	158	24 17	56	158	337
22	28 20	56	157	27 24	56	158	26 28	57	158	25 31	56	159	24 35	56	159	338
21	28 39	56	158	27 43	57	159	26 46	56	159	25 50	57	159	24 53	57	160	339
20	28 57	−57	159	28 00	−56	160	27 04	−57	160	26 07	−57	160	25 10	−57	161	340
19	29 15	57	160	28 18	57	161	27 21	58	161	26 23	57	161	25 26	57	162	341
18	29 31	57	161	28 34	57	162	27 37	58	162	26 39	57	162	25 42	58	163	342
17	29 47	58	162	28 49	57	163	27 52	58	163	26 54	58	163	25 56	57	164	343
16	30 02	58	163	29 04	58	164	28 06	58	164	27 08	58	164	26 10	58	164	344
15	30 16	−58	164	29 18	−58	165	28 20	−58	165	27 22	−59	165	26 23	−58	165	345
14	30 29	58	165	29 31	58	166	28 33	59	166	27 34	58	166	26 36	59	166	346
13	30 42	59	166	29 43	59	167	28 44	58	167	27 46	59	167	26 47	58	167	347
12	30 53	59	167	29 54	58	168	28 56	59	168	27 57	59	168	26 58	59	168	348
11	31 04	59	168	30 05	59	169	29 06	59	169	28 07	59	169	27 08	59	169	349
10	31 13	−59	169	30 14	−59	170	29 15	−59	170	28 16	−59	170	27 17	−59	170	350
9	31 22	59	170	30 23	59	171	29 24	60	171	28 24	59	171	27 25	59	171	351
8	31 30	59	172	30 31	60	172	29 31	59	172	28 32	60	172	27 32	59	172	352
7	31 37	59	173	30 38	60	173	29 38	60	173	28 38	59	173	27 39	60	173	353
6	31 43	59	174	30 44	60	174	29 44	60	174	28 44	60	174	27 44	59	174	354
5	31 48	−59	175	30 49	−60	175	29 49	−60	175	28 49	−60	175	27 49	−60	175	355
4	31 53	60	176	30 53	60	176	29 53	60	176	28 53	60	176	27 53	60	176	356
3	31 56	60	177	30 56	60	177	29 56	60	177	28 56	60	177	27 56	60	177	357
2	31 58	60	178	30 58	60	178	29 58	60	178	28 58	60	178	27 58	60	178	358
1	32 00	60	179	31 00	60	179	30 00	60	179	29 00	60	179	28 00	60	179	359
0	32 00	−60	180	31 00	−60	180	30 00	−60	180	29 00	−60	180	28 00	−60	180	360

Addendum 7 HO249 Vol II page 203

Lat 33° Declination 15°-29° Contrary

LHA ♈	Hc	Zn	Hc	Zn	Hc	Zn	Hc	Zn	Hc	Zn	Hc	Zn	Hc	Zn
	•VEGA		Alphecca		ARCTURUS		•SPICA		REGULUS		•POLLUX		Dubhe	
180	13 48	052	43 46	083	57 03	106	41 38	152	56 47	238	35 28	283	59 40	347
181	14 28	053	44 36	083	57 51	107	42 02	153	56 04	239	34 39	284	59 28	346
182	15 08	053	45 26	084	58 39	108	42 24	154	55 20	240	33 50	284	59 15	345
183	15 48	054	46 16	084	59 27	109	42 45	156	54 37	241	33 01	284	59 02	344
184	16 29	054	47 07	085	60 14	110	43 05	157	53 52	242	32 12	285	58 47	343
185	17 10	054	47 57	085	61 01	111	43 25	158	53 08	243	31 24	285	58 33	343
186	17 51	055	48 47	086	61 48	112	43 43	159	52 22	244	30 35	286	58 17	342
187	18 32	055	49 37	086	62 35	113	44 00	161	51 37	245	29 47	286	58 01	341
188	19 14	056	50 27	087	63 21	115	44 16	162	50 51	246	28 58	286	57 44	340
189	19 55	056	51 17	087	64 06	116	44 31	163	50 05	247	28 10	287	57 27	340
190	20 37	056	52 08	088	64 51	117	44 45	165	49 18	248	27 22	287	57 09	339
191	21 19	057	52 58	088	65 36	118	44 57	166	48 31	249	26 34	288	56 51	338
192	22 01	057	53 48	089	66 20	120	45 09	167	47 44	250	25 46	288	56 32	338
193	22 44	057	54 39	089	67 03	121	45 19	169	46 57	251	24 58	288	56 13	337
194	23 26	058	55 29	090	67 46	123	45 28	170	46 09	252	24 11	289	55 53	336
	•Kochab		VEGA		Rasalhague		•ANTARES		SPICA		•REGULUS		Dubhe	
195	46 32	011	24 09	058	24 48	091	12 46	134	45 36	172	45 21	252	55 32	336
196	46 41	010	24 52	059	25 38	091	13 22	134	45 43	173	44 33	253	55 12	335
197	46 49	010	25 35	059	26 28	092	13 58	135	45 48	174	43 45	254	54 50	335
198	46 58	010	26 18	059	27 19	092	14 33	136	45 53	176	42 57	255	54 29	334
199	47 06	009	27 01	060	28 09	093	15 08	136	45 56	177	42 08	255	54 06	334
200	47 14	009	27 45	060	28 59	094	15 43	137	45 58	179	41 19	256	53 44	333
201	47 22	009	28 28	060	29 49	094	16 17	138	45 58	180	40 30	257	53 21	333
202	47 29	008	29 12	061	30 39	095	16 50	139	45 57	181	39 41	258	52 58	332
203	47 36	008	29 56	061	31 30	095	17 23	139	45 56	183	38 52	258	52 34	332
204	47 43	007	30 40	061	32 20	096	17 56	140	45 52	184	38 03	259	52 10	331
205	47 49	007	31 24	061	33 10	097	18 28	141	45 48	186	37 13	260	51 46	331
206	47 55	007	32 08	062	34 00	097	18 59	142	45 42	187	36 24	260	51 22	331
207	48 01	006	32 53	062	34 49	098	19 30	142	45 36	188	35 34	261	50 57	330
208	48 06	006	33 37	062	35 39	098	20 01	143	45 28	190	34 44	262	50 32	330
209	48 11	006	34 22	063	36 29	099	20 31	144	45 18	191	33 54	262	50 07	330
	•Kochab		VEGA		Rasalhague		•ANTARES		SPICA		•REGULUS		Dubhe	
210	48 16	005	35 06	063	37 19	100	21 00	145	45 08	193	33 04	263	49 41	329
211	48 20	005	35 51	063	38 08	100	21 29	146	44 56	194	32 15	263	49 15	329
212	48 24	004	36 36	063	38 58	101	21 57	146	44 44	195	31 24	264	48 49	329
213	48 28	004	37 21	064	39 47	102	22 24	147	44 30	197	30 34	265	48 23	328
214	48 31	004	38 06	064	40 36	102	22 51	148	44 15	198	29 44	265	47 57	328
215	48 34	003	38 52	064	41 25	103	23 18	149	43 59	199	28 54	266	47 30	328
216	48 36	003	39 37	064	42 14	104	23 43	150	43 42	201	28 04	266	47 03	328
217	48 39	002	40 22	065	43 03	105	24 08	151	43 23	202	27 14	267	46 36	328
218	48 41	002	41 08	065	43 52	105	24 33	152	43 04	203	26 23	268	46 09	327
219	48 42	002	41 54	065	44 40	106	24 56	152	42 44	204	25 33	268	45 42	327
220	48 43	001	42 39	065	45 28	107	25 19	153	42 22	206	24 43	269	45 15	327
221	48 44	001	43 25	066	46 16	108	25 41	154	42 00	207	23 53	269	44 47	327
222	48 44	000	44 11	066	47 04	108	26 03	155	41 37	208	23 02	270	44 20	327
223	48 45	000	44 57	066	47 52	109	26 24	156	41 13	209	22 12	270	43 52	327
224	48 44	359	45 43	066	48 39	110	26 44	157	40 48	210	21 22	271	43 24	326
	DENEB		VEGA		•ALTAIR		ANTARES		•SPICA		Denebola		•Dubhe	
225	25 53	051	46 29	066	19 31	092	27 03	158	40 22	212	42 53	259	42 56	326
226	26 32	052	47 15	067	20 21	093	27 22	159	39 55	213	42 04	260	42 29	326
227	27 11	052	48 01	067	21 11	093	27 40	160	39 27	214	41 14	260	42 01	326
228	27 51	052	48 48	067	22 02	094	27 57	161	38 59	215	40 24	261	41 33	326
229	28 31	052	49 34	067	22 52	094	28 13	162	38 29	216	39 35	262	41 04	326
230	29 11	053	50 20	067	23 42	095	28 28	163	37 59	217	38 45	262	40 36	326
231	29 51	053	51 07	067	24 32	096	28 43	164	37 29	218	37 55	263	40 08	326
232	30 31	053	51 53	068	25 22	096	28 56	165	36 57	219	37 05	264	39 40	326
233	31 11	053	52 40	068	26 12	097	29 09	166	36 25	220	36 15	264	39 12	326
234	31 52	054	53 26	068	27 02	097	29 21	167	35 52	221	35 25	265	38 43	326
235	32 33	054	54 13	068	27 52	098	29 32	168	35 18	222	34 35	265	38 15	326
236	33 13	054	55 00	068	28 42	098	29 43	169	34 44	223	33 44	266	37 47	326
237	33 54	054	55 46	068	29 32	099	29 52	170	34 09	224	32 54	267	37 18	326
238	34 35	055	56 33	068	30 21	100	30 01	171	33 34	225	32 04	267	36 50	326
239	35 16	055	57 20	068	31 11	100	30 08	172	32 58	226	31 14	268	36 22	326
	•DENEB		ALTAIR		Nunki		•ANTARES		SPICA		ARCTURUS		•Alkaid	
240	35 57	055	32 00	101	17 43	140	30 15	173	32 22	227	62 47	246	60 22	314
241	36 39	055	32 50	102	18 15	140	30 21	174	31 44	228	62 00	248	59 46	313
242	37 20	055	33 39	102	18 47	141	30 26	175	31 07	229	61 14	249	59 09	313
243	38 01	056	34 28	103	19 18	142	30 30	176	30 29	230	60 27	250	58 32	312
244	38 43	056	35 17	104	19 49	143	30 33	177	29 50	231	59 39	251	57 55	312
245	39 24	056	36 06	104	20 19	144	30 36	178	29 11	231	58 52	252	57 17	312
246	40 06	056	36 54	105	20 49	144	30 37	179	28 31	232	58 04	253	56 40	311
247	40 48	056	37 43	106	21 18	145	30 37	180	27 51	233	57 15	254	56 02	311
248	41 30	056	38 31	106	21 46	146	30 37	181	27 11	234	56 27	255	55 24	311
249	42 12	056	39 19	107	22 14	147	30 35	182	26 30	235	55 38	255	54 46	311
250	42 54	057	40 07	108	22 41	148	30 33	183	25 49	235	54 50	256	54 07	310
251	43 36	057	40 55	109	23 08	148	30 30	184	25 07	236	54 01	257	53 29	310
252	44 18	057	41 43	110	23 34	149	30 26	185	24 25	237	53 12	258	52 50	310
253	45 00	057	42 30	110	23 59	150	30 21	186	23 43	238	52 22	259	52 12	310
254	45 42	057	43 17	111	24 24	151	30 15	187	23 00	238	51 33	259	51 33	310
	•DENEB		ALTAIR		Nunki		•ANTARES		ARCTURUS		•Alkaid		Kochab	
255	46 24	057	44 04	112	24 48	152	30 08	188	50 43	260	50 54	310	45 46	348
256	47 06	057	44 50	113	25 11	153	30 00	189	49 54	261	50 16	309	45 35	348
257	47 49	057	45 37	114	25 34	154	29 52	190	49 04	262	49 37	309	45 25	347
258	48 31	057	46 22	115	25 56	155	29 42	191	48 14	262	48 58	309	45 14	347
259	49 13	057	47 08	116	26 17	156	29 32	192	47 24	263	48 19	309	45 02	347
260	49 55	057	47 53	117	26 37	157	29 21	193	46 34	264	47 40	309	44 51	347
261	50 38	057	48 38	117	26 57	157	29 08	194	45 44	264	47 01	309	44 39	346
262	51 20	057	49 22	118	27 16	158	28 55	195	44 54	265	46 22	309	44 27	346
263	52 02	057	50 07	120	27 34	159	28 42	196	44 04	265	45 43	309	44 15	346
264	52 44	057	50 50	121	27 52	160	28 27	197	43 14	266	45 04	309	44 02	346
265	53 27	057	51 33	122	28 08	161	28 12	198	42 24	267	44 25	309	43 50	345
266	54 09	057	52 16	123	28 24	162	27 55	199	41 33	267	43 46	309	43 37	345
267	54 51	057	52 58	124	28 39	163	27 38	200	40 43	268	43 07	309	43 24	345
268	55 33	057	53 39	125	28 53	164	27 21	201	39 53	268	42 28	309	43 11	345
269	56 15	057	54 20	126	29 06	165	27 02	202	39 03	269	41 49	309	42 57	344

LAT 33°N

LHA ♈	Hc	Zn	Hc	Zn	Hc	Zn	Hc	Zn	Hc	Zn	Hc	Zn	Hc	Zn
	•DENEB		ALTAIR		•Nunki		ANTARES		ARCTURUS		•Alkaid		Kochab	
270	56 57	056	55 00	128	29 19	166	26 43	203	38 12	269	41 10	309	42 44	344
271	57 39	056	55 40	129	29 30	167	26 23	204	37 22	270	40 31	310	42 30	344
272	58 21	056	56 19	130	29 41	168	26 02	205	36 32	271	39 53	310	42 16	344
273	59 02	056	56 57	132	29 51	169	25 40	206	35 41	271	39 14	310	42 02	344
274	59 44	056	57 34	133	30 00	170	25 18	207	34 51	272	38 35	310	41 48	343
275	60 25	055	58 10	134	30 08	171	24 55	208	34 01	272	37 57	310	41 33	343
276	61 07	055	58 46	136	30 15	172	24 31	209	33 10	273	37 18	310	41 19	343
277	61 48	055	59 20	138	30 21	173	24 07	209	32 20	273	36 40	310	41 04	343
278	62 29	054	59 54	139	30 27	174	23 42	210	31 30	274	36 01	310	40 49	343
279	63 10	054	60 26	141	30 31	175	23 16	211	30 40	274	35 23	311	40 34	343
280	63 50	053	60 57	143	30 35	176	22 50	212	29 50	275	34 45	311	40 19	343
281	64 30	053	61 27	144	30 38	177	22 23	213	28 59	275	34 07	311	40 04	342
282	65 10	052	61 56	146	30 39	179	21 55	214	28 09	276	33 29	311	39 49	342
283	65 50	052	62 23	148	30 40	180	21 27	214	27 19	276	32 51	311	39 34	342
284	66 29	051	62 49	150	30 40	181	20 58	215	26 29	277	32 13	311	39 18	342
	•Alpheratz		Enif		ALTAIR		•ANTARES		ARCTURUS		•Alkaid		Kochab	
285	25 34	071	45 56	112	63 14	152	20 29	216	25 39	277	31 35	312	39 03	342
286	26 22	071	46 43	113	63 37	154	19 59	217	24 49	278	30 58	312	38 47	342
287	27 09	072	47 29	114	63 58	156	19 28	218	24 00	278	30 20	312	38 31	342
288	27 57	072	48 14	115	64 18	158	18 57	218	23 10	279	29 43	312	38 16	342
289	28 45	072	49 00	116	64 35	160	18 26	219	22 20	279	29 06	312	38 00	342
290	29 33	073	49 45	117	64 51	163	17 54	220	21 30	280	28 29	313	37 44	342
291	30 21	073	50 30	118	65 05	165	17 21	221	20 41	280	27 52	313	37 28	341
292	31 09	074	51 14	119	65 18	167	16 48	221	19 51	281	27 15	313	37 12	341
293	31 58	074	51 57	120	65 28	170	16 14	222	19 02	281	26 38	313	36 56	341
294	32 46	074	52 41	121	65 36	172	15 40	223	18 13	282	26 02	314	36 40	341
295	33 35	075	53 24	122	65 42	174	15 06	224	17 23	282	25 26	314	36 24	341
296	34 23	075	54 06	124	65 46	177	14 31	224	16 34	283	24 49	314	36 08	341
297	35 12	076	54 47	125	65 48	179	13 56	225	15 45	283	24 14	315	35 51	341
298	36 01	076	55 28	126	65 48	181	13 20	226	14 56	284	23 38	315	35 35	341
299	36 50	076	56 09	127	65 45	184	12 43	226	14 07	284	23 02	315	35 19	341
	•Mirfak		Alpheratz		•FOMALHAUT		ALTAIR		Rasalhague		•Alphecca		Kochab	
300	13 00	038	37 39	077	14 39	141	65 41	186	50 54	247	32 53	283	35 03	341
301	13 31	039	38 28	077	15 10	142	65 34	189	50 08	248	32 04	283	34 46	341
302	14 03	039	39 17	078	15 41	143	65 26	191	49 21	249	31 15	284	34 30	341
303	14 35	039	40 06	078	16 11	144	65 15	193	48 34	250	30 26	284	34 14	341
304	15 07	040	40 55	078	16 40	144	65 03	196	47 46	251	29 37	284	33 57	341
305	15 39	040	41 45	079	17 09	145	64 48	198	46 59	252	28 49	285	33 41	341
306	16 12	041	42 34	079	17 38	146	64 32	200	46 11	252	28 00	285	33 25	341
307	16 45	041	43 23	080	18 06	147	64 13	202	45 23	253	27 12	286	33 09	341
308	17 18	041	44 13	080	18 33	147	63 53	204	44 35	254	26 23	286	32 53	341
309	17 51	042	45 02	080	19 00	148	63 32	206	43 46	255	25 35	287	32 36	341
310	18 25	042	45 52	081	19 26	149	63 09	209	42 57	256	24 47	287	32 20	341
311	18 59	042	46 42	081	19 52	150	62 44	210	42 09	256	23 59	287	32 04	341
312	19 33	043	47 32	082	20 17	151	62 18	212	41 20	257	23 11	288	31 48	341
313	20 07	043	48 21	082	20 42	151	61 50	214	40 31	258	22 23	288	31 32	341
314	20 41	043	49 11	082	21 05	152	61 21	216	39 41	258	21 35	289	31 16	341
	•Mirfak		Hamal		Diphda		•FOMALHAUT		ALTAIR		•VEGA		Kochab	
315	21 16	044	23 21	076	16 19	125	21 28	153	60 51	218	60 29	292	31 00	342
316	21 51	044	24 10	077	17 00	126	21 51	154	60 19	220	59 42	292	30 44	342
317	22 26	044	24 59	077	17 41	127	22 13	155	59 47	221	58 55	292	30 28	342
318	23 01	045	25 48	078	18 21	127	22 34	156	59 13	223	58 09	292	30 13	342
319	23 36	045	26 37	078	19 01	128	22 54	156	58 38	224	57 22	292	29 57	342
320	24 12	045	27 27	079	19 40	129	23 14	157	58 03	226	56 35	292	29 41	342
321	24 47	045	28 16	079	20 20	130	23 33	158	57 26	227	55 48	292	29 26	342
322	25 23	046	29 06	080	20 58	130	23 51	159	56 49	229	55 02	292	29 10	342
323	25 59	046	29 55	080	21 36	131	24 09	160	56 10	230	54 15	292	28 55	342
324	26 36	046	30 45	081	22 14	132	24 26	161	55 31	231	53 28	292	28 40	342
325	27 12	046	31 34	081	22 51	133	24 42	162	54 52	233	52 42	292	28 24	342
326	27 48	047	32 24	082	23 28	133	24 57	163	54 11	234	51 55	292	28 09	343
327	28 25	047	33 14	082	24 05	134	25 12	164	53 30	235	51 09	293	27 54	343
328	29 02	047	34 04	082	24 40	135	25 26	165	52 49	236	50 22	293	27 39	343
329	29 39	047	34 54	083	25 16	136	25 39	166	52 06	237	49 36	293	27 24	343
	•Mirfak		Hamal		Diphda		•FOMALHAUT		ALTAIR		•VEGA		Kochab	
330	30 16	048	35 44	083	25 51	137	25 51	166	51 24	239	48 50	293	27 10	343
331	30 53	048	36 34	084	26 25	137	26 02	167	50 41	240	48 03	293	26 55	343
332	31 30	048	37 24	084	26 59	138	26 13	168	49 57	241	47 17	293	26 41	343
333	32 08	048	38 14	085	27 32	139	26 23	169	49 13	242	46 31	294	26 26	343
334	32 45	048	39 04	085	28 04	140	26 31	170	48 28	243	45 45	294	26 12	344
335	33 23	049	39 54	086	28 36	141	2[illegible] 40	171	47 43	244	44 59	294	25 58	344
336	34 01	049	40 44	086	29 08	142	26 47	172	46 58	245	44 13	294	25 44	344
337	34 39	049	41 35	087	29 39	143	26 53	173	46 12	246	43 27	294	25 30	344
338	35 17	049	42 25	087	30 09	144	26 59	174	45 26	246	42 41	295	25 16	344
339	35 55	049	43 15	088	30 38	145	27 04	175	44 40	247	41 56	295	25 03	344
340	36 33	049	44 05	088	31 07	146	27 08	176	43 54	248	41 10	295	24 49	345
341	37 11	049	44 56	089	31 35	147	27 11	177	43 07	249	40 24	295	24 36	345
342	37 49	050	45 46	089	32 02	148	27 13	178	42 20	250	39 39	296	24 23	345
343	38 28	050	46 36	090	32 29	149	27 14	179	41 32	251	38 54	296	24 10	345
344	39 06	050	47 27	090	32 55	150	27 15	180	40 45	251	38 08	296	23 57	345
	•CAPELLA		ALDEBARAN		Diphda		•FOMALHAUT		ALTAIR		•VEGA		Kochab	
345	20 44	048	14 06	079	33 20	151	27 14	181	39 57	252	37 23	296	23 44	345
346	21 21	048	14 56	080	33 44	152	27 13	182	39 09	253	36 38	297	23 32	346
347	21 59	049	15 45	080	34 08	153	27 11	183	38 21	254	35 53	297	23 20	346
348	22 36	049	16 35	081	34 30	154	27 08	184	37 32	254	35 08	297	23 07	346
349	23 14	049	17 25	081	34 52	155	27 04	185	36 44	255	34 24	297	22 55	346
350	23 52	049	18 14	082	35 13	156	27 00	186	35 55	256	33 39	298	22 43	346
351	24 31	050	19 04	082	35 34	157	26 54	187	35 06	256	32 55	298	22 32	347
352	25 09	050	19 54	083	35 53	158	26 48	188	34 17	257	32 10	298	22 20	347
353	25 48	050	20 44	083	36 11	159	26 41	189	33 28	258	31 26	299	22 09	347
354	26 27	051	21 34	084	36 29	160	26 33	190	32 39	258	30 42	299	21 58	347
355	27 06	051	22 24	084	36 45	161	26 24	191	31 49	259	29 58	299	21 47	348
356	27 45	051	23 14	085	37 01	163	26 14	192	31 00	260	29 14	299	21 36	348
357	28 24	051	24 04	085	37 15	164	26 04	192	30 10	260	28 30	300	21 25	348
358	29 04	052	24 55	086	37 29	165	25 52	193	29 21	261	27 47	300	21 15	348
359	29 43	052	25 45	086	37 41	166	25 40	194	28 31	262	27 03	300	21 05	348

Addendum 9 HO249 Vol I page 95

Lat 33° N Epoch 1975

TABLE 5.—Precession and Nutation Correction

L.H.A. ♈	North latitudes N.89°	N.80°	N.70°	N.60°	N.50°	N.40°	N.20°	0°	South latitudes S.20°	S.40°	S.50°	S.60°	S.70°	S.80°	S.89°	L.H.A. ♈
								1972								
°	mi. °	mi. °	mi. °	mi. °	mi. °	mi. °	mi. °	mi. °	mi. °	mi. °	mi. °	mi. °	mi. °	mi. °	mi. °	°
0	1 190	1 210	1 220	1 230	1 240	2 240	2 250	2 250	2 250	1 240	1 240	1 230	1 220	1 200	1 180	0
30	1 220	1 230	1 240	1 240	2 250	2 250	2 250	2 250	2 250	1 240	1 230	1 220	1 200	1 170	1 150	30
60	1 250	1 250	1 260	1 260	2 260	2 260	2 260	2 260	1 260	1 250	1 240	0 —	0 —	0 —	1 120	60
90	1 280	1 270	1 270	1 270	2 270	2 270	2 270	2 270	1 270	1 270	1 180	0 —	0 —	0 —	1 090	90
120	1 300	1 300	1 290	1 290	2 290	2 280	2 280	2 280	1 290	1 300	1 300	1 320	0 —	1 040	1 050	120
150	1 330	1 320	1 310	1 300	1 300	2 290	2 290	2 290	2 290	1 300	1 310	1 320	1 340	1 000	1 020	150
180	1 000	1 340	1 330	1 310	1 310	1 300	2 300	2 290	2 290	2 300	1 300	1 310	1 320	1 330	1 350	180
210	1 030	1 010	1 340	1 320	1 310	1 300	2 290	2 290	2 290	2 290	2 290	1 300	1 300	1 310	1 320	210
240	1 060	0 —	0 —	0 —	1 300	1 290	1 280	2 280	2 280	2 280	2 280	1 280	1 290	1 290	1 290	240
270	1 100	0 —	0 —	0 —	1 260	1 270	1 270	2 270	2 270	2 270	2 270	1 270	1 270	1 270	1 270	270
300	1 130	1 150	0 —	1 220	1 240	1 250	1 250	2 260	2 260	2 260	2 260	1 250	1 250	1 250	1 240	300
330	1 160	1 180	1 200	1 220	1 230	1 240	2 250	2 250	2 250	2 250	1 240	1 240	1 230	1 220	1 210	330
360	1 190	1 210	1 220	1 230	1 240	2 240	2 250	2 250	2 250	1 240	1 240	1 230	1 220	1 200	1 180	360
								1973								
0	0 —	0 —	0 —	1 230	1 240	1 240	1 250	1 250	1 250	1 240	1 240	1 230	0 —	0 —	0 —	0
30	0 —	0 —	1 240	1 240	1 250	1 250	1 250	1 250	1 250	1 240	1 230	0 —	0 —	0 —	0 —	30
60	0 —	1 250	1 250	1 260	1 260	1 260	1 260	1 260	1 260	0 —	0 —	0 —	0 —	0 —	0 —	60
90	0 —	1 270	1 270	1 270	1 270	1 270	1 270	1 270	1 270	0 —	0 —	0 —	0 —	0 —	0 —	90
120	0 —	1 290	1 290	1 290	1 280	1 280	1 280	1 280	1 290	1 290	0 —	0 —	0 —	0 —	0 —	120
150	0 —	0 —	1 310	1 300	1 300	1 290	1 290	1 290	1 290	1 300	1 310	0 —	0 —	0 —	0 —	150
180	0 —	0 —	0 —	1 310	1 300	1 300	1 300	1 290	1 300	1 300	1 300	1 310	0 —	0 —	0 —	180
210	0 —	0 —	0 —	0 —	1 310	1 300	1 290	1 290	1 290	1 290	1 290	1 300	1 300	0 —	0 —	210
240	0 —	0 —	0 —	0 —	0 —	0 —	1 280	1 280	1 280	1 280	1 280	1 280	1 290	1 290	0 —	240
270	0 —	0 —	0 —	0 —	0 —	0 —	1 270	1 270	1 270	1 270	1 270	1 270	1 270	1 270	0 —	270
300	0 —	0 —	0 —	0 —	0 —	1 250	1 250	1 260	1 260	1 260	1 260	1 250	1 250	1 250	0 —	300
330	0 —	0 —	0 —	0 —	1 230	1 240	1 250	1 250	1 250	1 250	1 240	1 240	1 230	0 —	0 —	330
360	0 —	0 —	0 —	1 230	1 240	1 240	1 250	1 250	1 250	1 240	1 240	1 230	0 —	0 —	0 —	360
								There is no correction for 1974.								
								1975								
0	0 —	0 —	0 —	0 —	1 060	1 060	1 070	1 070	1 060	0 —	0 —	0 —	0 —	0 —	0 —	0
30	0 —	0 —	0 —	1 070	1 070	1 070	1 070	1 070	1 070	0 —	0 —	0 —	0 —	0 —	0 —	30
60	0 —	0 —	0 —	1 080	1 080	1 090	1 090	1 080	0 —	0 —	0 —	0 —	0 —	0 —	0 —	60
90	0 —	0 —	0 —	1 100	1 100	1 100	1 100	1 100	0 —	0 —	0 —	0 —	0 —	0 —	0 —	90
120	0 —	0 —	0 —	1 110	1 110	1 110	1 110	1 110	1 110	0 —	0 —	0 —	0 —	0 —	0 —	120
150	0 —	0 —	0 —	0 —	1 120	1 120	1 120	1 110	1 120	1 120	0 —	0 —	0 —	0 —	0 —	150
180	0 —	0 —	0 —	0 —	0 —	0 —	1 120	1 110	1 110	1 120	1 120	0 —	0 —	0 —	0 —	180
210	0 —	0 —	0 —	0 —	0 —	0 —	1 110	1 110	1 110	1 110	1 110	1 110	0 —	0 —	0 —	210
240	0 —	0 —	0 —	0 —	0 —	0 —	0 —	1 100	1 100	1 100	1 100	1 100	0 —	0 —	0 —	240
270	0 —	0 —	0 —	0 —	0 —	0 —	0 —	1 080	1 080	1 080	1 080	1 080	0 —	0 —	0 —	270
300	0 —	0 —	0 —	0 —	0 —	0 —	1 070	1 070	1 070	1 070	1 070	1 070	0 —	0 —	0 —	300
330	0 —	0 —	0 —	0 —	0 —	1 060	1 060	1 070	1 070	1 060	1 060	0 —	0 —	0 —	0 —	330
360	0 —	0 —	0 —	0 —	1 060	1 060	1 070	1 070	1 060	0 —	0 —	0 —	0 —	0 —	0 —	360

The above table gives the correction to be applied to a position line or a fix, *deduced from the tables in this volume,* for the effects of precession and nutation. Each entry consists of a group of four figures of which the first is the distance in nautical miles, which the position line or fix is to be moved, and the group of three figures is the direction (true bearing). The table is entered first by the year, then by choosing the column nearest the latitude and finally the entry nearest the LHA ♈ of observation; no interpolation is necessary.

Example.—In 1972 a fix is obtained in latitude 23°N, when LHA ♈ is 71°. Entering the table with the year 1972, latitude 20°N, and LHA ♈ 60° gives 2 mi. 260° which indicates that the fix is to be transferred 2 miles in true bearing 260°.

Addendum 10 HO249 Vol I page 322

Table 5 - Precession and Nutation Correction

IN SUMMARY 21

You know enough about celestial navigation now to sail accurately wherever your little ship will go. It is hoped that this volume has taken the confusion and mystery out of the process. But, as with anything, there is no substitute for practice. You must actually do the work yourself. As a start in building your skill as a navigator, work through the examples carefully. Read the "Explanations" in the Air Almanac, and the "Preface" and "Introduction" to Volumes I and II of HO249.

Perhaps the best way of all to become familiar with the mechanics of celestial navigation is to take a simulated cruise and work out the positions and sights in reverse. Determine on paper first your DR position, time, and your Hc. Then work "backwards" to your sextant reading Hs and do your plotting. It is more work than the real process "forward," but the discipline is helpful.

You are now able to read and understand Dutton's and Bowditch. You will learn additional shortcuts from other navigators. Many of the niceties of navigation are now at your command. You can combine a visual or radio "line of position" with a celestial LOP for a "fix." A bearing on a distant mountain peak, or the cloud buildup over it, combined with a "sun sight," will give you an approximate "fix" when nothing else but the endless sea is visible. Don't worry about pin-point accuracy. That is not vital to safe landfalls.

Celestial Navigation **IS** fun, and a satisfying and valuable adjunct to your knowledge of ships and the sea.

GLOSSARY

ADVANCED LINE OF POSITION: A line of position which has been moved forward to allow for the run of the vessel since the line of position was established.

AIR ALMANAC: A publication sold by the U.S. Government Printing Office which tabulates astronomical data, covering in three volumes each year the periods of January through April; May through August; and September through December.

ALTITUDE: The angular distance of a celestial body above the horizon.

ARGUMENT: One of the values used for entering a sight reduction table.

ARIES: The position of the vernal equinox.

ASSUMED POSITION (AP): A point at which a vessel is assumed to be located. Used primarily to obtain entering arguments in sight reduction tables so that the computed altitude can be determined in the solution of a celestial observation.

AUTUMNAL EQUINOX: The point of intersection of the ecliptic and the celestial equator occupied by the sun as it moves from north to south declination on September 23.

AWAY: Line of position is away from an observed celestial body, and away from the assumed position.

AZIMUTH (Zn): Azimuth angle measured from the north pole clockwise through 360°.

AZIMUTH ANGLE (Z): The angular direction at the observer's position of a celestial object from the nearest pole measured through 180° East or West.

CELESTIAL COORDINATES: A set of coordinates used to define a point on the celestial sphere.

CELESTIAL EQUATOR: The intersection of the celestial sphere and the extended plane of the equator of the earth.

CELESTIAL POLES: The intersection of the celestial sphere and the extended axis of the earth.

CELESTIAL SPHERE: An imaginary sphere of infinite radius on which all celestial bodies, except the earth, are imagined to be located.

CIRCLE OF POSITION: A circular line of position.

COMPASS (Drafting): An instrument for drawing circles.

DAYLIGHT SAVING TIME: A change of time to the standard time of the time zone adjoining to the east. One hour later than standard time.

DEAD RECKONING (DR): Determination of position by advancing a previous position by reference to course and distance run.

DECLINATION: Angular distance north or south of the celestial equator measured north or south from 0° through 90°.

DIP: The vertical angle, at the eye of the observer, between the horizontal plane and the line of sight to the visible horizon.

DIVIDERS: An instrument consisting of two pointed legs, joined by a pivot, used for measuring distances on a chart.

ECLIPTIC: The apparent path of the sun as it moves around the earth on the celestial sphere.

EQUATOR: The primary great circle of the earth perpendicular to the polar axis.

FIX: A position determined from celestial observations, or other means, without reference to a previous position.

GEOGRAPHICAL POSITION (GP): That point on earth at which a celestial body is at the zenith at a specified time.

GREAT CIRCLE: The intersection of a sphere and a plane passing through its center.

GREENWICH HOUR ANGLE (GHA): Angular distance west of the Greenwich Meridian measured through 360°.

GREENWICH MEAN TIME (GMT): Local mean time at the Greenwich Meridian.

GREENWICH MERIDIAN: The meridian passing through Greenwich, England, serving as the prime meridian.

H.O. 249: Sight Reduction Tables published by the Defense Mapping Agency Hydrographic Center in three volumes and sold by the U.S. Government Printing Office. Volume I includes pre-computed altitudes and azimuths of selected stars; Volume II, for latitudes 0° - 39°, and Volume III, for latitudes 40° - 89°, provide complete data for the reduction of sun, moon, planet, and star sights within the declination range of 0° - 29°.

HEIGHT: The angular distance of a celestial body above the horizon.

HORIZON SYSTEM: Altitude and azimuth of celestial bodies based upon the observer's horizon.

HOUR ANGLE: Angular distance west of a celestial meridian measured through 360°.

HOUR CIRCLE: A great circle on the celestial sphere passing through the celestial poles.

INDEX CORRECTION: A correction to eliminate the error in the reading of a sextant equal to the difference between the zero position of the scale and the zero position of the instrument.

INTERNATIONAL DATE LINE: A hypothetical line approximately at the 180° meridian, where each calendar day first begins.

INTERCEPT: Altitude difference between observed and calculated altitudes expressed in nautical miles.

LATITUDE: Angular distance north or south of the equator measured from 0° through 90°.

LIMB: The circular outer edge of a celestial body, the top being the upper limb, the bottom being the lower limb.

LINE OF POSITION (LOP): A line on which a vessel is presumed to be located as a result of a celestial observation.

LOCAL APPARENT NOON: The instant at which the sun is on the local meridian.

LOCAL HOUR ANGLE (LHA): The angular distance of a celestial body west of the local celestial meridian measured through 360°.

LOCAL MERIDIAN: The meridian passing through the observer.

LONGITUDE: The angular distance east or west of the Greenwich Meridian measured from 0° through 180°.

MERIDIAN: A great circle passing through the North and South Poles of the earth.

NADIR: A point on the celestial sphere vertically below the observer.

NAUTICAL MILE: A distance of 6,076 feet, equal to the length of 1′ of latitude on the surface of the earth.

NAVIGATIONAL TRIANGLE: The spherical triangle used in calculating celestial navigation problems.

NOON SIGHT: A measurement of the altitude of the sun when it crosses the local meridian, used to obtain the latitude.

NUTATION: Irregularities in the precessional motion.

PARALLAX: The difference in the apparent direction of an object when viewed from different positions.

PLOTTER: An instrument used for plotting lines and measuring angles.

PLOTTING CHART: A chart designed for plotting lines of position from celestial observations.

POLARIS SIGHT: A measurement of the altitude of the pole star Polaris, used to obtain the latitude.

PRECESSION: Change in the direction of the axis of rotation of a spinning body.

PRIME MERIDIAN: The meridian of longitude 0°, the Greenwich Meridian.

PROTRACTOR: A drafting instrument used for measuring angles.

REFRACTION: The deflection from a straight line of a ray of light as it passes obliquely from a rarer to a denser layer of air.

ROUND OFF: To change a figure to the nearest whole minute of arc, or one nautical mile.

SEMI-DIAMETER: The angular distance between the limb of the sun or moon and the center of the body.

SEXTANT: An instrument for measuring angles, primarily altitudes of celestial bodies.

SEXTANT CORRECTION: A correction for an inaccuracy of a sextant due to mechanical or optical imperfections.

SIDEREAL HOUR ANGLE (SHA): Angular distance west of the vernal equinox (Aries), measured from 0° through 360°.

SIGHT: Observation of the altitude of a celestial body to determine a line of position.

SIGHT REDUCTION TABLE: A table containing computed altitudes and azimuths representing solutions of the navigational triangle required in obtaining a line of position.

STANDARD TIME: The civil time established by law over a specified time zone.

SUMMER SOLSTICE: A point on the ecliptic occupied by the sun on June 21 at maximum northerly declination.

TIME DIAGRAM: A diagram, as viewed from the south celestial pole, in which the celestial equator appears as a circle and celestial meridians and hour circles appear as radial lines.

TIME SIGNAL: A radio signal marking a specified time.

TIME ZONE: An area in all parts of which the same time is kept. The centers of adjoining time zones are approximately 15° apart in longitude.

TOWARD: Line of position is between assumed position and toward an observed celestial body.

TRUE BEARING: Bearing relative to true north.

VERNAL EQUINOX: The point of intersection of the ecliptic and the celestial equator occupied by the sun as it moves from south to north declination on March 20.

WINTER SOLSTICE: A point on the ecliptic occupied by the sun on December 22 at maximum southerly declination.

WORK SHEET: A printed form used in reducing a navigator's observations in establishing a line of position.

ZENITH: The point on the celestial sphere vertically overhead.

ZENITH DISTANCE (ZD). The angular distance of a celestial body from the zenith.

INDEX OF FIGURES